ALIVE
...The Missing Link

Kenjin

Yeti

Ape Rakshasa Chemosit

Rambai Urayuli Metch-Kangmi

Khün Hairy Shaitan

Skunk Bitchi Peevi

Sandja Beruang Dev

M'toto

Geresun Mo Kaftar Sogpa Almasti

Bambursh Jungli-Admi

Osodrashin Görüessü Almas

Snally-Gaster

Abnauayu

Mokololut Rakshi-Bompo

Man Zerleg

Harrum-Mo

Kühn Wulgaru

Kiik-Adam

Chuchuna

By Odette Tchernine

Typeset by Guinevere Palmer,
Layout by Muhammed and Ibrahim for CFZ Communications
Using Microsoft Word 2000, Microsoft Publisher 2000, Adobe Photoshop CS.

First published in Great Britain by CFZ Press

CFZ Press
Myrtle Cottage
Woolsery
Bideford
North Devon

EX39 5QR

© CFZ MMXXIII

Cover artwork 'ORANG PENDEK' by Steve Baxter
https://mysteriesillustrated.art/work

ISBN: 978-1-909488-69-4

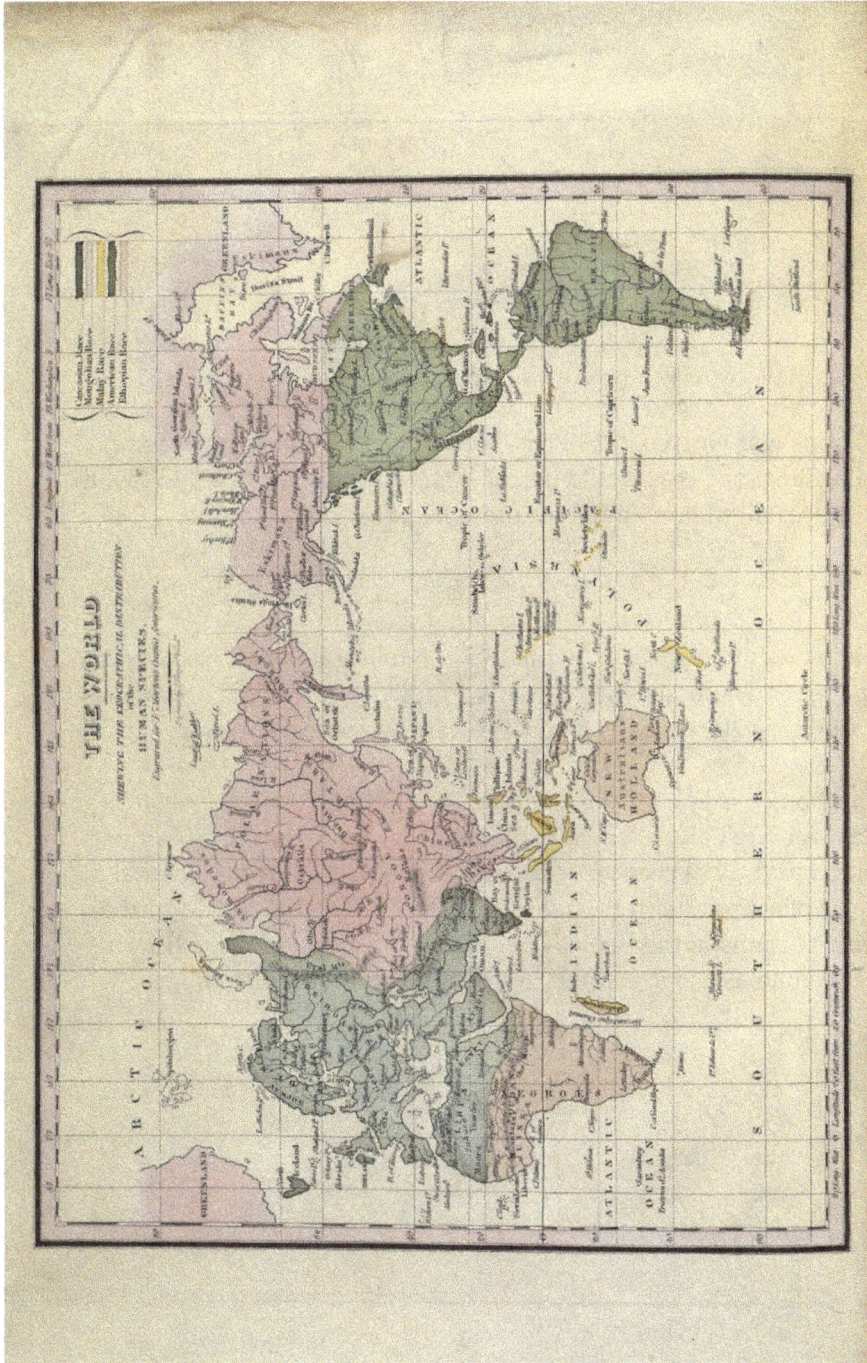

Odette's 'S Shaped map' will have covered from the Caucasus to Mongolia, a fringe of the Tibet-China borders, Siberia, and the Himalaya. Then over the Bering Straits through Alaska and down to British Columbia and Northern California. The original map is sadly lost.

This book was written in 1986 but never published. It was discovered as a paper copy (possibly written on a typewriter) in an auction house in early 2021, nearly 30 years after the author's passing.

This book has been transcribed by Guin Palmer, Administrative Director for The Centre for Fortean Zoology. The introduction through to (and including) chapter seventeen were written by the original author. Following chapters have been added for clarity and to further explain sections of the book. Additional footnotes added by Richard Freeman, Zoological Director of Centre for Fortean Zoology.

FOREWORD

by
Colonel John Blashford-Snell, CBE FRSGS

Odette Tchernine, an acknowledged journalist, novelist and author keenly interested in cryptozoology, was in her late seventies when I met her in London. She approached me at a lecture I gave on the crossing of the infamous Darien Gap, and from her bulging shopping bag produced various manuscripts for me to read about the Yeti. It was the result of some fascinating research into this mysterious subject, and Odette encouraged me to undertake investigations into the unexplained sightings during my expeditions which I did, largely in Nepal, Tibet and Mongolia, without being able to reach a firm conclusion.

I learned that this enthusiastic lady had been born in Paris in 1897, the daughter of a Russian financier and a French mother. Like many accomplished journalists she had numerous friends in Fleet Street, and it was thanks to her that I published my first autobiography.

Whenever we met she had new stories to tell me about the Yeti, and her research into reports and many anecdotal accounts. Odette was especially keen to discover if the alleged creature was a species of ostensibly extant hominin, or an animal. In spite of being obsessed by the subject she was still open-minded and up to the end of her days, eager to learn the truth behind the legend.

Hopefully this intriguing new book will encourage more research and protection of any creature that might be discovered.

John Blashford-Snell
Motcombe 2023

CONTENTS

The following chapters were added in 2023 by Guin Palmer,
Assistant Director for The Centre for Fortean Zoology:

INTRODUCTION

This introduction is based on the rapid strides the natural sciences are making to discover the truth of man's beginnings. My new approach (already foreshadowed in parts of my previous books) will lead on to chapters containing direct and indirect clues to humanity's start.

Many of the clues postulated by others, as well as mine, have hither-to been ignored. The uniformed consider them preposterous if considered at all. But among the informed some lines of scientific thought have begun to be considered seriously.

Suggested clues I refer to occupy no slot in anthropology's records, past or present.

Practically no serious investigation has been applied to the long existing phenomenon of unexplained hominid reports, even though fringe science research has gone on for centuries. Sporadic knowledge has on the whole only provided scope for the jokes and hoaxes. These have created disbelief in the long-standing rumours of what Rudyard Kipling once called "Something lost behind the ranges."

I have called this untapped branch of the tree of life "Anthropo-zoology."

The only serious public investigation mounted was when in 1978 the University of British Columbia organised a conference on the subject and its related folk memory spheres in Vancouver which I attended as listed observer. Its proceedings will be amplified in one of the "clue" chapters which will follow this introduction[1].

In spite of America's great advance in the sciences there are millions there who reject Darwin. Many of these Fundamentalists do not belong only to uninformed communities but are members of modern enclaves of society.

It can be dangerous to enlightenment when some of the Book of Genesis is taken too literally. Several minds produced the Bible nonetheless, and many truths are voiced in it. Some indicate hidden mysteries of earth's beginnings. The Inadmissible Possible - perhaps?

Of all the persecutions man has inflicted upon man down the ages, perhaps in the ultimate the most important might be the little-known hounding and near

---

[1] RF - The USSR had an official Snowman Commission in the 1950s based at the Darwin Museum in Moscow. They took an expedition to the Pamir Mountain and a number to the Caucasus. In 1977 the Chinese Academy of Sciences lanced a large expedition to search for the yeren.

destruction of groups of hominids who possibly from the accidental handicaps of possessing both human and animal characteristics, lost their way up a branch of life they should have ascended. By man's intolerance of something incomprehensible these creatures were hunted cruelly because they were different.

And yet by cunning born of extreme danger from fellow creatures, they learned how to vanish from the tribes and survive in small, scattered groups hidden in unapproachable reaches of the earth's remotest wildernesses.

The stories, myths and names, where this mystery belongs are found in the terrain of man's urge to bury uncomfortable rumours and evidence.

This has been achieved through legendary tales, and the true traditionally handed down names of **Yeti, Snowman, Metch-Kangmi, Migo, Bigfoot**, Sasquatch, Almas/ Almasty, Kaftar, Dev, Hairy Man, Kiik-Adam, Chemosit, Chuchuna. They are found not only in traditional Himalayan spheres, but in all earth's continents.

In my first book on the phenomenon, I thought the unidentified being was a large anthropoid. Gradually in the course of extended research and study I reached the conclusion that the identity was not so simple. Now I am convinced the mystery is more complex because of its double nature.

Man's fear of revelation could be the reason for the long disguise of terms applied to the mystery.

Pierre Teilhard de Chardin[2], the great priest-scientist spoke of God Himself being in the heart of all matter. This is true, and the Yeti/Snowman/Hairy Man mystery lies in that heart. It may be hidden there for a one transcendental purpose. The all-creative spirit is not vested in only one living being that has become Homo sapiens.

It is the living force in all nature.

Primitive societies that have not yet reached the metal age still exist. They are valuable to discovery. Relations between linguistic and cultural anthropology are necessary, but not yet familiarised enough among those who search. The anthropologist must know terms, languages of expression. In years gone by Hairy Man or unknown animal references caused errors from the misunderstanding of terms. Unexpected glimpses in the field brought similar mistaken interpretations

---

[2] https://en.wikipedia.org/wiki/Pierre_Teilhard_de_Chardin

through confusion of dialects, myths, superstition. It has been said that Sociology is the twin of anthropology. To this can be added that the cultural, political, historical and geographical spheres are then Anthropology's cousins. There is a continuous and widespread link.

Ancient heritages of knowledge and questioning on mankind were voiced by historians and philosophers of various terrains; Mediterranean China, mediaeval Arab centres. Some former contributors to this search were Jean Bodin[3], in the 16th century, Rousseau[4], and Voltaire[5]. In England's 19th century, protagonists included Thomas Hobbes[6] and John Locke[7], and in 1859 came Charles Darwin's "Origin of Species." Among various sources of interest Sir James Fraser's "Golden Bough," (1890)[8] a vast record of customs, religions, and magical practices, gained wide attention. Ruth Benedict[9] and Margaret Mead[10] now come to mind, and a prominent name in France of current interest is Claude Levi-Strauss[11].

The last century saw extensive supporters of the "Great Chain of Being". There followed the search for the Missing Link. At first this search only gave scope to entertainers, and showmen like Barnum who soon produced several little-known types of creatures which he presented to circus audiences as various missing links. It became a popular joke.

The Great Chain still has that missing link in its continuity.

In 1699 a traveller, Edward Tyson, wrote a book called "Anatomy of a Pygmie"[12]. It compared the monkey or ape with man. His work was a landmark in research. Here is a passage from it:

"Since therefore in all respects the brain of a Pygmie (Chimpanzee) does so exactly resemble man's, there is no reason to think that agents do perform such and such actions because they are found with organs proper there unto: For then our Pygmie might be really a man."

This passage is found in a fuller form in one of my further chapters as an early clue, if

[3] https://en.wikipedia.org/wiki/Jean_Bodin
[4] https://en.wikipedia.org/wiki/Jean-Jacques_Rousseau
[5] https://en.wikipedia.org/wiki/Voltaire
[6] https://en.wikipedia.org/wiki/Thomas_Hobbes
[7] https://en.wikipedia.org/wiki/John_Locke
[8] https://en.wikipedia.org/wiki/The_Golden_Bough
[9] https://en.wikipedia.org/wiki/Ruth_Benedict
[10] https://en.wikipedia.org/wiki/Margaret_Mead
[11] https://en.wikipedia.org/wiki/Claude_L%C3%A9vi-Strauss
[12] https://en.wikipedia.org/wiki/Orang-Outang,_sive_Homo_Sylvestris

a rather questionable one.

The rest of this book does not present clues to the Missing Link in all chapters, but some of them do. Among them is the record of Merlin Hellener's encounter with a Bigfoot when she stayed at a forest summer youth camp in America in the mid-sixties. The story has never been told before, and it was Ms Hellener in the recent 1970s/80's who gave it to me personally. Next in the clue range comes the female Bigfoot film shot in a split second in Northern California in 1967.

The Chuchuna evidence of Arctic Siberia qualify for inclusion, as does the story of the Russian technical scientist woman, Serikova, who woke up in her holiday cottage in a remote part of a Moslem area of the U.S.S.R. to find an unprepossessing Almas creature squatted on the floor studying her curiously. She had never heard of such creatures before. Mountaineer Don Whillans'[13] 1970 photograph of footprints, and his sighting at night on the heights of Annapurna also justify being mentioned.

There are others that deserve attention from established science that it has obviously not been possible yet for scientists to apply. Probably remnants of the Missing Link are alive, but terrains for proof are forbidding, such as the mountains of outer Mongolia, forests of Central China, even still the Himalayan first venue, and the Americas' wilder regions, global cooperation would be required, and huge financial outlay involved.

And what nations with vital hunger for oil and defence programmes, and with one hopes, uneasy consciences over some Third World's actual physical hunger zones, will spend a fortune in searching for these mysteries?

The riddle whom Mirabehn[14] (Madeleine Slade) in conversation with me, once called "Our lost brother of the mountains."

---

[13] https://en.wikipedia.org/wiki/Don_Whillans
[14] https://en.wikipedia.org/wiki/Mirabehn

Chapter One
Setting the Pace

My two previous books on this subject. 'The Snowman and Company'[1] and 'The Yeti'[2] were the result of years of delving in between my other necessary writings. They were the result of my search and of the co-operation I received and still receive from serious fact hunters like me. The delving has had to be carried out side by side with practical digs in other fields - literary and newspaper work - the hard stuff needed for the immediate living and for the caring for others.

These two books go on ambling their way in Britain, other countries, and within university precincts: in the Americas, where the Yeti's equivalent, Bigfoot and Sasquatch, keep rearing shaggy heads in that continent's "forests of the night", and impressing extraordinary footprints here and there in soil or sand. In Russia my delving seems to have even reached Siberia! Rather unexpected and surprising since several countries always possessed records of the phenomenon, scattered about. Could this so-called legendary and hairy creature be laughing his head off when he places his huge flat foot on soil that may already and unfortunately have been marked by the occasional hoaxers who fake tracks and have confused the issue for years by repelling serious scientific approach?

But science is gradually coming round, and researchers continue with field work and laboratory tests on the sparse evidence available. Doubtful information is sifted. Researchers make permanent notes of the newest reports that stand up to examination, and they keep in touch with one another in this Snowman saga which I have grown to call by the title of Anthropo-Zoology.

Data goes back to the earth's early centuries, ranging sporadically from the Himalaya, China, Mongolia, remote Russian areas of the Caucasus Mountains, America, Canada, and Africa. Scattered clues occur at widely separated spots.

Researchers are dispersed, but we travel when means allow. We communicate, and exchange news, whether we be laymen "who want to know" or professors and faculties from Oxford to Ulan Bator. It is a worldwide informal network.

The Yeti has been walking again, rarely, unexpectedly, and in fresh quarters. In subsequent pages will appear new unpublished reports from the earth's four

---

[1] https://www.amazon.co.uk/Snowman-company-Odette-Tchernine/dp/B0000CKZEU
[2] https://www.abebooks.co.uk/book-search/title/the-yeti/author/odette-tchernine/

corners. Presented too will be a few debateable stories, bandied about at times by those with a misplaced sense of humour rather than a sense for accuracy. There is no malice in my recording a few discredited tales together with the convincing and sometimes personally experienced ones. The nonsense is quoted deliberately as examples of the jokes that can handicap genuine exploration.

I shall try to dig back into some of the accepted records of mankind's long-vanished past because they are coincidental with factors existing in the present Yeti situation. We shall refer to certain of the progresses in the natural sciences, and some comparatively modern discoveries about man's or near-man's, evolving. Attention will at times focus on some deviations in the life stream. These will be reviewed because in each case quoted there could be an indirect association with the Abominable Snowman riddle.

There is a strange and persistent image. It owes its name in the English language to the discovery in 1921 by Colonel C.K. Howard-Bury[3] of mystery footprints of large though clumsy human appearance, that he found on the Himalayan Lhakpa-La Pass. This was 21,000 feet in altitude northeast of Everest. The expedition's porters, as they called the local guides in those days, said the footprints were the tracks of a Metch-Kangmi[4]. Kang means snow, and Mi, man. Mi presented no problem, but Metch did.

Climbers, experts, and the scholars took a tumble over Metch. It seemed to derive from Tibetan sources and/or some Himalayan vernacular. It suggested, dirt, raggedness, and a disgusting appearance. So the term 'Abominable Snowman' was born together with the controversy about the creature's existence.

Other areas have given the phenomenon other names according to geographical habitat, ranging across two continents, and possibly more. These include the Asian mountains of India, China, Tibet, the Wildernesses of Russia and the Americas. Such names and regions will be dealt with later on in this, my third record of a pursuit.

In documentation of this nature, it is not always possible to treat it entirely chronologically. Now and again, it is necessary to re-cap, or dodge back and forth in time. This is because some new light is shed, or the allegation of a hitherto little-known distribution is added to the mystery. Previous known factors are extended following fresh evidence or theory. Or ancient stories that carry convincing

---

[3] http://anomalyinfo.com/Stories/1921-colonel-howard-burys-sighting
[4] This should read 'metoh-kangmi', (a Sherpa word meaning man of the rocks) rather than 'metch-kangmi'.

comparisons turn up halfway through the business of writing a chapter. Again, a second visit to some distant area may provide fresh aspects to an older story.

During investigations I have at times changed my first tentative opinions because of such new data and personal communication with fellow researchers. This would be an attempt to apply a flexible approach, and not to hold entirely to any early belief if events suggest it could have been a mistaken assessment.

Exchange of growing knowledge is the only means towards solving a problem. Especially in a search in Anthropo-Zoology.

There is now widening understanding that the sciences of life upon earth are not bound in separate spheres of knowledge, but relate to one another in various degrees, and even at times merge. This merging is illustrated in Yeti research, mine and that of past distinguished researchers, often silenced so as to avoid embarrassment.

In 1978 the "Sasquatch and Similar Phenomena" Conference[5] was mounted for the first time in centuries. An attempt to throw more light on the mystery. This happened from May 10th-13th at the University of British Columbia in Vancouver about the phenomenon of The Yeti, Abominable Snowman, Bigfoot, Metch-Kangmi, Almasty, Kaftar, etc. The last three being some of the names the Russians have given for years to their own Anthropo-Zoological mysteries.

The four-day conference was the first on the subject ever to be treated at serious academic level. This was chiefly due to the pioneer work and foresight of the university's Professor Marjorie Halpin[6], curator of the University's anthropological museum, whom I had already met in 1976 on a previous visit. That time we had a fascinating and useful conversation during which the Yeti/Snowman question was analysed.

I was listed as an observer to their 1978 conference[7]. It presented a mixed bag of academics, scientists, researchers, hunters, laymen enthusiasts and critics. Some fur did fly, but a good informative time was had by all who attended. The exercise's central figure was of course absent while the four-day debates revolved around him/her/it. The conference presented the phenomenon's many varying characteristics and those of its possibly near-relatives and the general stamping grounds.

---

[5] https://sasquatchchronicles.com/sasquatch-the-anthropology-of-the-unknown-1978-part-3/
[6] https://en.wikipedia.org/wiki/Marjorie_Halpin
[7] https://atom.moa.ubc.ca/index.php/conference-participants-addresses

Brought to the huge lecture hall was the controversial Female Bigfoot film,[8] the late Roger Patterson and his companion, Bob Gimlin took in 1967 at Bluff Creek, Northern California. Many of us present had seen it then and had puzzled over it repeatedly since it first burst upon a goggling world of science and disbelief. It was later tested in depth and pronounced genuine by the Russian researchers of the Darwin Museum, Moscow - Dmitri Bayanov[9], Igor Burtsev[10], Igor Donsky and other colleagues. This was when René Dahinden[11] took it to Moscow in the early seventies before Professor Boris Porshnev died. Dahinden has been one of the few dedicated Bigfoot investigators for almost a lifetime, and even if he thinks the riddle will not be solved until someone shoots a Yeti and brings back the body for analysis, I consider he is genuine, though I am against killing.

The Russian group of colleagues in our research follow in the footsteps of Boris Porshnev. While being well known as a historian in Russia, Porshnev spent nearly a lifetime investigating the Yeti mystery. He travelled extensively in known areas of "Relict Hominid" distribution and produced resulting documentary works. These have since proved invaluable to authors and researchers including me on the phenomenon, information which he provided generously.

Dmitri Bayanov and his colleagues were unable to attend the meetings in person, but his paper analysing the film was read by Professor Marjorie Halpin, and she read Jeanne Koffman's[12] paper too (a detailed account of the film will be given further on). Marjorie Halpin then read a paper of her own. This has related importance and concerned her studies of the monkey/ape masks created by some American-Indian tribes, and the ritual significance of these strange artefacts. Masks of malevolent Influence for some, or of beneficial character for others. There were indirect links with some aspects of the Yeti mystery. Jeanne Koffman's contribution covered the field work in the Caucasus she and her team of assistants has done for many years. Some of such details have been published before[13]. These Almasty were described as shy semi-human hairy creatures of forests and other remote wildernesses, raiding isolated orchards or dairy settlements, and living and breeding harmlessly and animal-like in desolate hide-outs.

Several years ago, the Almasty were more numerous than now, and were accepted locally as part of the Caucasus's more savage regions. If encountered, they were

---

[8] https://en.wikipedia.org/wiki/Patterson%E2%80%93Gimlin_film
[9] https://www.thebigfootportal.com/dmitri-bayanov/
[10] https://www.pinterest.co.uk/pin/327918416594990950/
[11] https://en.wikipedia.org/wiki/Ren%C3%A9_Dahinden
[12] https://www.wikidata.org/wiki/Q18043824
[13] https://www.isu.edu/media/libraries/rhi/research-papers/Koffmann_1.pdf

treated kindly. Almasty babies looked like ordinary human infants, and only became hair-covered and developed other unusual physical characteristics later. The adult Almasty had bigger jaws and teeth than had humans, and possessed abnormally long arms, hanging below the knees. This trait is associated with Yeti and other descriptions of the phenomenon where distribution has been reported and, in some cases, observed. Here, one could compare the Almasty babies' developing to body hairiness with the discoveries of two French doctors of more than a century ago. These two Drs Le Double and Houssay, made a lifelong study of hairy human freaks of the world. This led to their unique book, "Les Velus"[14] ("The Hairy Ones"). Their works will be related in a later chapter.

The Vancouver conference heard several speakers whose papers were not of direct Sasquatch nature, but which could be associated to the mystery in the fields of folklore, superstition, and racial rites and beliefs. There is a great deal of demon fear among some communities. At one time, some Indians declared that Satan and his brood would soon attain world power. Some of the conference audience might perhaps have thought this idea could be currently applicable. Anthropologist Julius Kassovic[15] of the University of California told how a remote Mexican village made a cult of supplying pictures and specimen artefact demons for the edification of tourists. Some of the artists employed in this strange promotion work would frighten themselves with their own creations and would hide them at night so as not to be scared to death by what they had perpetrated. Nearby Indian communities called such exhibitions the haunts of devils and monsters. These cults may be of value to the tourist trade but are not so beneficial to natives. One wonders if such side-tracks into fantasies help in serious Yeti research. I doubt it.

Peter Byrne[16] of the Bigfoot Research Museum in Oregon was present. So were Nicholas and Diane Webster, who earlier in 1978 had produced a documentary Yeti film for the Alan Landsburg Film Company[17] of Los Angeles. Part of it was produced in London and several of persons prominent in Yeti research took part, including me.

Then came the conference's highlight moment when the Female Bigfoot movie was shown. It was projected from every angle: Right, left, centre, backwards, forward, slow-motion, and stationary for the viewer's benefit.

---

[14] https://www.abebooks.co.uk/velus-Contribution-l%C3%A9tude-variations-exc%C3%A8s-syst%C3%A8me/8002269663/bd
[15] https://archives.dickinson.edu/image-archive-people/kassovic-julius-s
[16] http://www.petercbyrne.com/greatsearches.html
[17] https://en.wikipedia.org/wiki/Alan_Landsburg_Productions

Many sceptics of the 1967/8 film showing admit now that it may be genuine. If, as the Russians state, it is no hoax, though still unidentifiable in nature. When René Dahinden had taken it to Moscow, it was tested frame by frame, and the projections examined piecemeal. Factors tested were season of the year, time of day, light and shade, nature of soil, and weight calculations based on the central figure's movements and progress.

When first seen in London in 1968, like many other viewers, I had an open mind, as to what it purported to show, the first moving photography of the Bigfoot phenomenon, though only a brief exposure. Now I incline to believe it is genuine, though possible the biggest mystery of the Yeti saga.

For those who may not know the full history of the Female Bigfoot movie, here it is. The film was shot by the late Roger Patterson and his companion, Bob Gimlin, at Bluff Creek, Northern California, in October 1967. Incidentally, Gimlin was at the Conference and sounded quite convincing.

The Bluff Creek encounter seemed to have been unexpected, though Patterson and his friend had been in the bigfoot search in likely places before this incident.

So, on came the "shock" movie: The wooded mountainside, broken boughs and logs lying along the forest verge. Humanly impenetrable brush everywhere except on the rough logger's track. The camera suddenly caught movement, focused around wildly as it picked on the huge, black, hair-covered creature emerging upright from the trees. It charged across the scene, beast or semi-human, heavy pressure in movement and step, enormous, fantastic. The head turned, ape-like, looked for a split time-fraction into the lens, face showing palish colour. The creature vanished back into the brush from which it had appeared. Every time one sees this film one strains to pick out some thing new, a clue, anything, but each time (and I had seen it several times before) the mind and eyes cannot decide.

Dahinden described his years-long search for Bigfoot, and of more recent years, his striving to prove if the film is a hoax or genuine. He is still unconvinced one way or the other. Dmitri Bayanov and colleagues nonetheless believe that Patterson's Bluff Creek movie is genuine.

In Bayanov's own words read by Professor Halpin, the film is "A triumph of broadmindedness against narrow-mindedness." Bayanov considers the creature is neither ape nor man, but states that it passed the tests they applied to the film. The shots were stated to have been taken 40 metres from the camera. The creature's gait and the way the feet dug into the soil were conclusive. Bayanov said that a human hoaxer in a fur suit would never show such a gait, nor the way the creature's muscles flexed under that black fur. The Russians calculated its weight to have been at least 220 kilogrammes, and the height about 200 centimetres.

Grover Krantz[18], anthropologist professor at Washington State University, was present at the conference. He is one of the few academics who researched consistently on the Sasquatch footprints and handprints. He has produced papers on this part of his work. But he admits that the sceptics have so far wanted a Bigfoot body to prove its existence and later in this book I shall give his work and views a more detailed presentation.

How often are corpses of known animal species ever found in their natural habitat? Seldom. Climate and predators soon dispose of dead flesh and bones.

To kill or not to kill is a controversial question among genuine researchers. Some agree with Krantz that acquiring a body of a shot Sasquatch will be the only means of clearing up the mystery. Opposing views hold that there must be no shooting, but only tranquillizing guns, difficult, perhaps impossible as such a method would be in such wild forested areas, especially when sightings are so rare and transient. For what my views are worth, I stand for non-killing.

Present too at the conference was John Green[19], until recently established at Harrison Hot Springs, British Columbia, where he has explored Bigfoot areas and collected evidence, and taken plaster casts of the many footprints he discovered. He has been a lifelong researcher in this sphere and has written several documentary books about it. When in 1976 at Harrison Hot springs myself on my first B.C. visit before I attended the Vancouver 1978 conference, I went up the Lillooet Mountains up beyond the Harrison Hot Springs village where he had an office-museum, and a collection of a thousand-odd plaster casts of footprints. We took photographs at a spot high up, driving on loggers' mountain tracks. We stopped on the mountain at a place overlooking Harrison Lake far below. We had pulled up to take pictures of the very clearing where a year or so previously there had allegedly been a Sasquatch sighting at very close quarters. This is described where I shall deal with other comparatively recent incidents of this nature.

---

[18] https://en.wikipedia.org/wiki/Grover_Krantz
[19] https://en.wikipedia.org/wiki/John_Willison_Green

Professor Carleton Coon[20], famous veteran American anthropologist, wound up the conference's final evening session. He had made an exhausting trip from his home in New Hampshire where he has retired. He gave a good talk and was not afraid to include some humour. He stressed that Sasquatch/Bigfoot might have been a Northern American immigrant from Asia during the Ice Age. The Bering Sea land bridge of those ancient times must have attracted quite a few strangers. He said Sasquatch could have been the earliest man-type inhabitants of North America. He suggested they might have been put there to survive and "give hominids a new chance". I already had that thought comparison. Coon, one of the famous old names in anthropology, was not afraid of reminding us with a smile how the human races seems bent on "finishing one another off".

The University of British Columbia Conference on "The Sasquatch and Similar Phenomena" has not produced a Yeti. This was not expected. But it brought together personalities of experience who have made the problem a serious study and continue to do so. It has presented a wealth of material, indicating that despite the hoaxes, nonsense stories, mistakes, and contradictions, there are substantial reasons for this enquiry to be continued and eventually placed on a world-wide basis.

Professor Marjorie Halpin will go down in the history of research and discovery as the first person who was able to take the first established step in this direction.

---

[20] https://en.wikipedia.org/wiki/Carleton_S._Coon

Chapter Two
On the Snows of Annapurna

My second book on the phenomenon, "The Yeti" was about to appear in 1970 when the British Expedition to climb Annapurna in Nepal returned to Britain after a successful ascent. So, one curious experience of one of its members, Don Whillans, could not be included in the book. Nor could the photograph he took of the mystery tracks. This was before he sighted something unusual and unidentifiable on the mountain slope during the night that followed his taking the picture in daytime.

That day of return in London, he and his teammates sat on the platform of their first press conference almost immediately after they had come back. I had spoken to him briefly just before it started, and he was questioned during the meeting about what he saw. A few days later he came to see me, and we discussed the incident more fully.

During the expedition which was to climb the south face of Annapurna, Whillans and Douglas Haston[1] had climbed ahead of the main team to establish an advance base in readiness for the final assault to reach the summit.

Whillans found and photographed a trail of footprints in the snow. A few hours later at night, from the temporary hut he and the Sherpas had set up, he saw a mystery living creature out on the slopes. The photograph he had taken during the day is the clearest and most definite picture of unidentifiable footprints since Eric Shipton's world-famous one in the Himalaya, whose story of discovery has often been told. This happened in 1951 on the Menlung Glacier when he and Dr. Michael Ward[2] were on the Everest Reconnaissance investigation. The photograph is still the most challenging indication of the Yeti riddle. The footprint dimensions are about 12 & ½ inches by roughly 6 & ½ inches width. These measurements rather tally with those of Don Whillans' tracks picture on Annapurna. Dimensions have varied in different parts of the world where similar geographical conditions obtain.

Don Whillans' footprint photograph is as puzzling as Shipton's was, though not as definite because his shows a series of those footprints progressing in diminishing size up a snowbound slope. But compared with Shipton's recording of the one

---

[1] https://en.wikipedia.org/wiki/Dougal_Haston
[2] https://www.alpinejournal.org.uk/Contents/Contents_1999_files/AJ%201999%2081-87%20Ward%20Footprints.pdf

solitary track, which for clarity was photographed apart from others on the glacier, Whillans shot taken in 1970, nineteen years after Shipton's, poses the same question.

When Whillans was questioned at the meeting, he spoke reluctantly and briefly of the unidentified shape he saw after he took the picture. When he came to see me a few days later we spoke more fully of this experience. He said that for half an hour, through binoculars, he had watched from the darkness of the hut, the black hairy shape moving with bi-pedal tread over the moonlit snow. It was zigzagging about, bending down sometimes, occasionally dropping on all fours as if searching for possible edible roots buried under the snow. It finally vanished in a thick belt of trees higher up on the mountain.

The altitude was roughly 14,000 feet, indicating terrain below the snowline in spite of the snow lying around. The actual recognized snowline is the altitude where it seldom melts. Heavy snowfalls can occur below this, however. Whillans said he had not tried to photograph the creature because at night he would have had to use flash which would have frightened away whatever it was, and so shortened time for observation. He had studied its movements and approximate size and shape for nearly half an hour of concentrated effort, but this could not give him a real clue of the visitant's nature. He added that this part of the mountain where the advance base was being pitched had not had human beings on it before. It was a hitherto unexplored region of Annapurna. The only living souls who had come there were now the witnesses of this sighting, Whillans and two Sherpas who were accompanying him and Haston. It is not clear if Haston was around at the time. I think not.

The Sherpas were in no doubt the tracks had been made by a Yeti, and that the creature they watched that night as it bounded about in the snow was the same creature that had made the footprints some hours earlier. Later, Whillans learned that this part of the Annapurna range had sometimes been known as "The Valley of the Great Ape."

He had never before taken any interest in the unexplained mystery of the Himalaya or of anywhere else where the same phenomenon was reputed to exist. He is not an anthropological or zoological student, so he was not looking out for what he saw or expecting anything abnormal. This lack of interest or anticipation give credibility to the incident, even though he must have received some chaff about it. Whillans had never bothered about the remnant existence of the Yeti as a species, but he did see what he saw, and proof of something unusual lies in his photograph of the footprints.

Sometime later, I had a friendly telephone chat with Audrey Whillans during her husband's absence again at the Himalaya. I was just getting some photographic information but talk veered around informally to the Annapurna incident. According to her, there had been no liquor around on that night that could have caused a little eyesight blurring to arise, followed later by jokey exchanges at the camps. And the altitude was not high enough to occasion hallucinations. I remember we got a bit of fun at this point, for she recalled that on that particular night no consignment of beer for the whole expedition had yet arrived!

The tracks photograph is real enough. Whillans' experience must have stimulated his curiosity. He told me how he had been puzzled at first by one of the tracks being more deeply impressed than the rest. In the lower section of the photographs one of the footprints is much deeper sunk than the others and has a different shape. Some time after his return from Annapurna, Whillans was reading a zoological book. It described how some of the big anthropoids when going down on all fours have a habit of leaning heavily on double-under front knuckles. An illustration in this book showed this very characteristic. It looked exactly like the bottom footprint in Don's tracks photograph. It must have been made when the creature went down on all fours the better to dig out something from under the snow.

Since Annapurna there have been several expeditions by various nationalities to the Himalayan regions, for different purposes, but generally basically to beat previous high-altitude records, or to scale hitherto unclimbed peaks.

The story of Don Whillans' Annapurna experience has received scant attention. Yet its basic feature is that his photograph is the most distinct known picture of unidentified footprints in the Himalaya since Eric Shipton's remarkable one taken on the Menlung Glacier in 1951.

The 1951 photograph was the one that caused such sensation among challenging scientists that one of them mounted an exhibition of stuffed Langur monkeys with imprints of their feet in sand. This was to suggest that Shipton and his companions had photographed a langur footprint.

In his genuine efforts to preserve the accepted belief that all fauna had been discovered and scientifically catalogued, the expert had been a little too hasty. The Langur's paws, or feet, have long toes, rather like the digits of a human hand, and their imprints in the artificially prepared soil were nothing like the shape of the foot in the Menlung Glacier photograph.

Surprising discoveries are often received and dismissed because to pursue them to a definite conclusion might cause embarrassment, though a few dedicated experts have continued pursuing controversial discoveries in the interest of extended or revised knowledge. The lack of subsequent interest in the Don Whillans photograph and statement, is a case in point of the fear of embarrassment. The same old line of dismissal is safer.

Among the different expeditions to the Himalaya during the last decades one Japanese trip to Everest in the late sixties caused mild if transient interest, though of a lighter nature. The Japanese dedication to photography sometimes amounts to addiction. Wherever they travel, they appear to photograph everything, bristling with cameras at the ready even on strenuous mountaineering in which it must be admitted they excel. One report about one of their trips percolated to the West.

Keen Japanese climbers had come across mystery footprints, they said, while engaged on an Everest exercise. They took a large number of photographs, most of which the West never saw, except for the later efforts of one persistent Briton. But this comes in later. Among the Everest Japanese mountaineers was an American who reported on mystery tracks several months after their expedition was over. He said the footprints were thought to be those of an Abominable Snowman/Yeti. When pressed for a description he replied that they were enormous shuffling marks. He added jokily that they looked as if they had been made by Donald Duck ploughing through the snow in carpet slippers. Then the jokey American vanished from publicity. The serious element in Japanese alpine activities no doubt dismissed such comments as too frivolous to expand across the world's media.

Other British quarters than the first who divulged this odd story were interested, and I was commissioned to find out more. This was during the 70's, and I underwent a cable and telephone saga back and forth between certain Japanese gentlemen and myself. Eventually I dug out some information, and some pictures. The information was in Japanese, and linguists had to help. It was a non-event, for the photographs only looked as if someone had been dragging parcels or parkas through the snow. Tracks galore, but not the right ones. Yes. I think I was the only person in the United Kingdom who obtained pictures of a sort.

---

[3] https://en.wikipedia.org/wiki/Kelvin_Kent_(mountaineer)

Major Kelvin Kent[3] had been a member of one British Everest expedition of the seventies that had had to be abandoned because of exceptionally severe weather. He sent me a print of some strange foot print photographs he had taken in the snow just as a curiosity. The picture showed criss-cross triangular tracks. They began and ended as if to and from nowhere. We agreed that they were merely the tracks of a largish bird that had alighted for a second and flown off. The claw marks were obvious. The American climber's joke about Donald Duck tracks in the snow was probably not so far-fetched after all!

But the Japanese are persevering. In 1974 they were reported to be searching for the Yeti again. A team of twelve women went to Kathmandu to climb a 26,650 foot peak in the central Himalaya, and to look for the Snowman, but no spectacular results happened. Later on, one solitary Japanese woman was said to have scaled Everest, but there are doubts.

The term "Yeti" seems to have had Nepalese connotations. Names change according to area. Around the map of Northern India, would be heard terms like Migo, Rakshasa, Rakshi-Bompo. There are others. Then on the Russian scenes we have the Almasty, Dev, Kaptar, Kiik-Adam, and Shaitan. Some are names used in more remote regions, and some belong to Moslem tradition.

Bigfoot is in the Americas as well as Sasquatch at the Canadian end. Chemosit, a doubtful starter, but known for long, occurs in East Africa. It was once said to denote the legendary Nandi Bear.

The Yeti has walked in many places, and there are more regions of equal geographic nature where traces of the phenomenon and its legends occur. In fact, now this has become more of an actuality, as continued reading will show.

In my research and study, my tracing the sightings, encounters, sounds and legends still remain within the areas of my S-Map which covers two continents. I gave the map this name because when traced on an atlas the areas of alleged distribution formed a giant letter S lying on its back.

Perhaps soon I may discover another alphabetical map outline by tracing another distribution, a little known one, dotted over part of another continent. But my old S-Map still holds good, even though many stories of Snowman sightings and sounds within this map's radius are often repetitions of what has already been told. Or often told inadmissible reports which were ultimately revealed not to be admissible at all, but had arisen either from a vivid imagination or from a case of mistaken identity. A case in point could be illustrated by the mistaken sight of some mountain solitary or eccentric, living and feeding with the yak herds during seasons when the higher pastures are favourable to grazing, and were spotted from a distance a human drop-out's wild and unkempt appearances gives rise to animal-like comparisons. Such isolated instances can occur sometimes but can certainly make no claim to snowmanship.

Quite apart from mountain eccentrics, some workers in the natural sciences continue their fauna studies and function in the field. An Arun Valley Wildlife expedition did valuable research in recent years. The group were engaged in studies in the Himalayan area lying between Everest and the Kanchenjunga peak. The Thailand Association for the Conservation of Wildlife co-operated, we learn, and three Britons were engaged in the Arun Valley project. There was also cooperation and advice from the Fauna Preservation Society of London, it is believed. The United Nations appear to have taken a sponsoring interest, while the Nepalese Foreign Office in Kathmandu and the British Embassy there gave useful advice to the expedition's principals. They were J.A. McNeeley, E.W. Cronin, and H.B. Emery.

One of their experiences provided a slight piece of Snowman evidence. On December 18th, 1973, Cronin and Emery's tent was on a ridge in the Upper Arun. During the night it was visited outside by an unseen animal. It left tracks that were not identifiable as belonging to any known species[4]. This kind of inconclusive evidence has occurred before during many decades in the Himalaya.

The mountaineer, the late Wilfred Noyce[5], was a member of Lord Hunt's team for the successful Everest ascent. He told me a good many years later of an experience he and some of his companions had one night, when in their hut during the acclimatising period before the final assault on the peak. Noyce happened to look out on the mountain. It was very dark, but though nothing living was visible, there was an impression that something had been prowling around the hut. It had uttered curious whistling sounds, whatever it was. The sound is characteristic, and the Sherpas had identified it as coming from a Yeti. Many explorers and naturalists have heard similar noises as those noted by Wilfrid, and saw puzzling footprints in snow or soil, as is well known. These witnesses are persons of such integrity that their words cannot be doubted. Apart from Noyce, from Lord Hunt, who has quoted similar experiences, to the late H.W. Tilman and the late Eric Shipton, is to quote a list of distinguished names.

---

[4] RF - Biologist and mammalogist George Schaller likened the tracks found around Cronin and Emery's tent to those of a mountain gorilla.
[5] https://en.wikipedia.org/wiki/Wilfrid_Noyce

One far older footprint witness can be added. He was quite an unknown traveller until I dug him out of the British Museum Library, or rather, dug out his forgotten book. In it, the author, H.J.B. Fraser, described an unpretentious and rather prosy Himalayan tour when he saw strange footprints at which his porters took fright and told him were "Bang's" occasionally one of the names attached to the Abominable Snowman. But nobody knew this in those days, for the book was published in 1820, one hundred years before Colonel Howard-Bury found tell-tale footprints on an Everest pass. I make no excuse for repeating this true story from my previous books - my discovery of the first Yeti reference published in the English language.

Dedicated investigators are treating the Yeti question as seriously as those scared porters did more than a century ago but are setting about it objectively.

In later chapters will be found recent American and British Columbia factors, which like the Himalayan and other searchings have puzzled experts and amateurs for years past down to the present day. Russian progress will also be mentioned, as well as contemporary revelations about another continent which I refer to earlier in my S - Map comments.

However, I will end this Chapter Two with the reported true story of the 1980's: About French-Canadian Trapper, Georges Gilot, rescued from an icefall by Yeti-type Hairy giants who saved his life.

George Gilot, a French-Canadian trapper in the Yukon, broke his leg in an ice fall near his camp. He lost consciousness. He awoke in a cave, his gear beside him. He had been rescued by a family of hairy, Yeti-type giants. They stared at him through huge, glittering eyes.

Terrified, he fainted again. He came to and found his broken leg had been packed with snow to try and heal it.

Gradually, he and his rescuers communicated in sign language - earth's earliest talk.

The father would tell Gilot when he went out for meat and water by making signs or mime. The days wore on in the extraordinary atmosphere. Gilot set his broken leg with splints from his tent pole.

The family communicated among themselves in grunts and groans.

There was the father, mother and a boy and girl - their children. They behaved tenderly and kindly to one another. When sleeping, they would huddle together like little puppies.

One day, Gilot awoke to find them gone and all his gear neatly packed beside him. He understood them to mean that he was now fit to fend for himself, and this he did. He collected his equipment and, coping with his badly set leg - he knew he could never be a trapper again as he made his way down the mountains to Dawson. From there he told he wanted to go to Saskatchewan and try to write a book. He wanted to write about his wonderful saviours, their strange world, and their almost thoughtful family life of which the outer world knew nothing.

This report came to me exclusively from the Yukon via a colleague in California who keeps me informed of out-of-the-ordinary news which often concerns my research and occasional travels investigating nature phenomena. The dateline of the Georges Gilot story was as recent as October 1983.

Here is the much older similar story I mentioned at first: When I heard it some twenty years ago, I thought it was a hoax. Now expert opinions of the wild far north Canadian areas consider it was true. It concerned Canadian geologist/mountaineer Albert Ostman[6], who was captured by a family of ugly hairy giants in the same altitudes as the Gilot story. They were not unkind, but they would not let him go until at last in romping with two hairy giant children, he made his escape. Afraid of being laughed at, he never told his story for years. Gilot's hairy rescuers make a happier picture, but there is a great "look-alike" thing about it. Indirectly to this, I think some of the American Indian tribes could tell us a lot about hairy yeti-type giants of their remote mountains.

---

[6] https://en.wikipedia.org/wiki/Albert_Ostman

Chapter Three
Backward Looks

The Abominable Snowman has an important entry in the Encyclopaedia Britannica in which the Yeti is described as a myth. Controversy and contradictions continue in this case just as it does concerning the Loch Ness underwater creatures.

Many years ago, in Northern India's Garhwal region, tribesman confronted and killed an alleged Snowman that had captured one of their women. They did not bring back its body from the high-altitude cave where they had rescued their unharmed kinswoman. Mirabehn (Madeleine Slade) to whom they were responsible as herdsmen, asked them why. They replied they had not dared bring back their kill as they might have been charged with murder. To them the "myth" of the High Himalaya was a very real wild hairy man dwelling in the almost impenetrable forests and caves below the snowline. A creature that occasionally descended to raid their crops and orchards when conditions were favourable and vanished when it felt itself observed.

At that time Mirabehn who was entirely "Indianised" after her becoming Mahatma Ghandi's devoted disciple, was managing a cattle improvement project for the Indian Government. She was in complete authority. Those Garhwali tribesmen's assessment of the nature of the Snowman equates with reports from other primitive communities in mountains and forest-grown terrains widely separated one from the other.

The full story of Mirabehn and her high-altitude cattle staff's adventure appears in my previous books on Yeti research[1].

A. N. Tombazi in 1925 was on forestry work in Sikkim. When his men early one morning called him to come and look at a Yeti foraging for roots among Rhododendrons, he probably could not believe his eyes when he saw the dark fur-covered thing a few yards down the mountainside in bright sunshine. Afterwards, he treated the incident as "one of those delicious Sherpa fairy tales".

Yet this official of long ago may have been the only European to have seen a Yeti at such close quarters.

---

[1] The Snowman and Company" (Robert Hale Ltd). "The Yeti"(Neville Spearman in U.K.) Published in USA by Ta-plinger, Inc., titled "In Pursuit of the Abominable Snowman".

J.R.C. Gent, another official, had a somewhat similar experience earlier in the century. His workmen in the timber belt near Phalut told him of a creature that had frightened them. They called it the Sogpa or the Jingli-Admi.

The Himalayan botanist and traveller of the last century, H.J. Elwes[2] F.R.S. spoke of these incidents some years later in 1915. He was something of a zoologist too, and he told how he had once seen a Yeti in Tibet and had sketched it and taken notes. His lecture on the subject was widely discussed. Royal Botanical Society members had seen his report and picture, and so had the Zoological Society. Unfortunately, sketch and report were never found when required many years later. I personally tooth-combed all information sources a few years ago when I was writing my first Yeti book, but like others before me, was unsuccessful. Sketch and notes had vanished for ever.

Sometimes one knows for certain that certain clues of long past facts did exist, and yet they can never be followed up, and it is a waste of time to try too long. Research progresses slowly and links in information can be lost. This creates the blanks and gaps in knowledge. This might be because various groups are often unaware of another's activities and step-by-step records. Professor Boris Porshnev whose lifelong Yeti research was carried on side by side with his academic work commented once that it was only in the 1950's that Russia learned of the existence of a similar Yeti question in the Himalaya. Now, Dmitri Bayanov and colleagues at the Darwin Museum, Moscow, continue Porshnev's work.

General communication has become highly organised and efficient, but specialised communication is the Cinderella of the media. Yet it should be in free reporting and exchange of specialised information that new knowledge grows.

For many years Tibetan monks and the Himalayan ones, were responsible for a large number of Yeti reports. Some were feasible, others far-fetched, though, pronounced by them in genuine belief they were right in promoting the Snowman image. The Compass (mountain temples) were alleged to possess real Yeti relics in their museums. These included the once controversial Snowman scalps. One was later proved to have been devised from the hind quarters of a goat[3]. Others too were artefacts. But here mystery arises, for hairs taken from one "scalp" proved when analysed to have belonged to some unknown species. This unknown thread of hair came from the scalp belonging to the Pangboche Temple.

---

[2] https://en.wikipedia.org/wiki/Henry_John_Elwes
[3] RF - Biologist and mammalogist George Schaller likened the tracks found around Cronin and Emery's tent to those of a mountain gorilla.

An exhibit skull from Khumjung village was lent to Sir Edmund Hillary to show it to Western lecture audiences on his return from one of his Himalayan expeditions. These relics are very ancient and surrounded by Gompa traditions handed down by the lamas' word-of-mouth and brought out to use as head-dresses during ceremonial rites. Except for their folklore connotations, they are of little use to Anthropology or Zoology. Still the unidentified hair sample remains and there are several collected at different times. Dr William C. Osman Hill[4], former Prosector of the London Zoo, at one time made a study of them. He was the author of "Abominable Snowman, The Present Position"[5], a short book report published by the Fauna Preservation Society some years ago during his time of office at the Zoo.

Other alleged evidence of Yeti existence was brought forward from time to time, and individual professional and amateur explorers took to the Yeti trail. One of them, an American Oil Millionaire, Tom Slick[6], was so dedicated to the search that he spent prodigally in briefing experienced hunters to follow the trail. Expense was no object. It is probably that Slick's enthusiasm led him up many blind alleys. He may even at times have been tricked by informants. In some cases, they produced 'genuine' relics that were not genuine at all.

One of these objects was actually believed to be the real thing by its finder. It was supposedly a mummified Yeti's hand. I saw the colour film of it at the London Zoological Society at the time when Slick's expeditions were at their height. One of his staff present seemed perturbed at my presence. They were showing the film in good faith, but it was discovered some time later that the rather gruesome colour photograph had been of a gorilla's withered hand. Slick died some years later without running to earth the Yeti on which he spent fortunes.

More than one century ago, long before Professor Boris Porshnev began his lifelong quest to uncover the mystery of Russia's own type of Snowman, Colonel Nikolai Przhevalsky[7] of the Imperial Army of Tsarist days, came up against the ignorance in some rare aspects of the natural sciences that he has trammelled research for centuries. He returned from explorations in remote Russia and Mongolia with the story of the Almasty, human-like animals, or animal-like humans that he had discovered. They were later to be described as unacceptable.

[4] https://en.wikipedia.org/wiki/William_Charles_Osman_Hill
[5] https://www.cambridge.org/core/services/aop-cambridge-core/content/view/DBCC8D9681D6CBE2930820D8744ABF3D/S0030605300001253a.pdf/abominable-snowmen-the-present-position.pdf
[6] https://en.wikipedia.org/wiki/Tom_Slick
[7] https://en.wikipedia.org/wiki/Nikolay_Przhevalsky

Russia had known nothing of these strange dwellers of the vast continent's wildernesses, and those in power did not want to know about any unspecified form of life occurring in their remote mountains and Central Asian deserts. The court of those days did not actually label him as a liar, but they disliked the nature of his explorations. These were hushed up as embarrassments and left hidden in archives until comparatively recent Russian authorities uncovered these documentaries about their native phenomenon and sponsored some investigation.

There were knowledge gaps in Przhevalsky 's day, but his research in depth was not the first-time enlightened seekers after truth had pressed to bring to the surface knowledge about the Almasty. Several learned professional and governmental men had spent all the time they could in travels and expeditions to relevant parts of the Russian Empire to solve the problem. Details of their discoveries, disappointments, and frustrations could be found in scattered papers, but I succeeded in assembling everything possible of centuries-long research in my two previous books. As this present one contains the newest general data I could gather, and also will contain my latest findings from several other parts of the world not touched upon before, the long past Russian data is only mentioned briefly. Yet this bizarre episode is worth repeating: Official modern Russia first sponsored an expedition to find the Snowman in the Pamir Mountains, and suddenly called it off. There was no such creature, and Hydrologist Dr. Pronin who had previously sighted a hairy person high up on an inaccessible rock face was treated derisively. The frown of officialdom continued for some years. The seventies with its advance of enquiries around the world brought a more moderate view in the Russian media sources. This must certainly have been due to the work of the Darwin Museum in Moscow. Though standing apart from the official Russian Academy of Sciences, the Baynov group there, is becoming more and more known.

A few years ago, I collected a few facts about a Russian woman's strange experience in 1956 while she was on holiday in Kabardino, a far-off and outlying area of the U.S.S.R. I now have the full story which could be an example of the puzzling nature in so-called Yeti types. N.Y. Serikova was a zoological technologist. So, the Russians described her profession. She was in Kabardino purely on a vacation as there was no mention in the Russian version of her following any course of study while she was there.

She was lodging in a collective farmer's home. One late afternoon she had been out in the countryside for a walk. The farmer's house was empty when she returned, and she remembered that the family were visiting neighbours where there had been a wedding and where a party was now taking place. Pleasantly tired after her walk in what was probably rather rough country, she lay down on her bed and half fell asleep.

She suddenly started up, awakened by a screech quite close at hand, and sat up quickly looking around.

Squatting on the floor of her room only a few paces away sat an incredibly ugly creature looking hard at her. Its left hand raised across to its right shoulder, and right one on left. It looked as if about to leap up and attack her. Almost paralyzed with fright, Serikova managed to cry out: "Oh Lord, where do you come from?"

She was at pains to explain later on that the exclamation when she had called upon the name of the Almighty was by accident and from sheer horror, as she was not a God-believer!

The thing on the floor gave another loud screech, and then jumping upright in shambling style, banged outside and into the next house, the whole cottage shaking from the impact. The creature left behind an unpleasant sour smell. Serikova was still shaken and frightened. She told later how she locked herself in, being alone, except for the unwelcome "guest" next door, and she was terrified it might come back. She dared not go out or move until next morning for - "I thought it was something devilish," she said.

Then she learned about the Almasty, and that this had been one of the species. It had once become a sort of tamed, incongruous pet of an old woman who lived in the next cottage and had died. After that the Almas (the singular term) had moved into the home where Serikova was staying, and the collective farmer and his family had just casually accepted this strange heritage.

The creature had been scared when it heard Serikova cry out in Russian, and not in the Kabardino dialect with which it (she or he) was familiar. Also, the Almas had been disturbed by the noise of the next-door wedding party and merry-making, and this had added to the creature's panic.

Local people could not tell Serikova any scientific facts about the Almasty. After she had left the region, she kept thinking of her extraordinary experience.

"It was the height of a medium-sized man," she told friends, "With its body covered with hair, not very long, but thick. It had heavy bushy eyebrows, black hair on its face with that hair shorter than on the body. It was an ugly face like an animal, and yet like a mans. The thing was only about one metre away from me when it jumped up with that awful scream and dashed away. That this was an Almas and not a real man, was clear now that I know about them. The shape of its eyes sloped down, not up, and from its 'wild beast' way of looking at me, I guessed even then that it was not a real human. And also, from its foul smell. The shape of the head was not a man's head, and it was somewhat elongated."

Serikova's mind was finally put at rest five years later when she learned that investigators from Moscow were in Kabardino to study the question of the Almasty.

Soon after her return from that holiday she appeared to have spoken to country people who knew about the creatures. She said: "I have often had conversation now about this with cattle breeders, and they told me they had seen Almasty, or heard about them from parents, grand parents, and friends. I spoke to ordinary people too, like shepherds and herdsmen. These people never lie when they trust you and are sure of your good faith. Folk are afraid of seeing Almasty. They get frightened of them through the Mullahs" - she was speaking of the Moslem regions "And they say with conviction that if they betray an Almas his kinsmen will wreak vengeance upon the offender."

This woman's description was very accurate and clarified after her enquiries following her holiday. Apart from excessive hairiness, the Almas had a very primitive clumsy body and brutish face, head and eyes suggestive more of a stone-age pre-Neanderthal remnant type than of the anthropoid ape with which the general idea of the Yeti has been compared.

Those strange unacceptable man-animals or animal-men of the Russian and Mongolian wilder areas do not quite tally with the Himalayan anthropoidal image. Is there justification in considering that the Almasty, that is, the Yeti-Snowman conception, to be similar to the Himalayan image, or as having affinity with the American Bigfoot, the British Columbia Sasquatch and its other Canadian names?

Are the Hairy Ones as brutish as their appearance suggests.

In some remote forests, a few persons who have come across one accidentally and for the briefest moment, have described their startled eyes as gentle and full of curiosity before they vanished again in the brush. These wild ones have never injured anyone, though accidents might have happened through mutual panic. They may not be savage, the few that remain, and may even have feelings enough to bury their dead.

When tracing a reason for unsolved side-tracks in evolution we find labels attached to this species or that which are not always as rock-sure as they purport to be.

During Palaeontology research in caves at Shanidar near Bagdad several years ago, a Neanderthal grave was excavated. Its significance of a spiritual nature was only realised many years later. There were signs of funeral rites. The body, that of a young man, had been laid on a bed of pine branches entwined with wildflowers. The enclosing sediment's condition in the deep soil had preserved the pollen. Care taken in these last rites showed that the rough Neanderthals were not the popularly believed primitive brutes without mind or sensitivity[8].

Clues to human and other living creatures' habitat and custom are often found when not searching for them, as the above account shows. Like an unusual paragraph on Page 57 of "King Jesus" by Robert Graves. Simon is describing Jerusalem as midway "between India and Spain - between the frozen White Sea of the North and the insufferably hot deserts to the Southward, where the Ape-Men beat devilishly upon their hairy chests, and East and West are confounded..."

Who were these so-called ape-men of so long ago?

---

[8] "Statement by Mme Leroi-Gourhan, Musee de l'Homme, Paris. Quoted by Science Editor Robert Chapman, Sunday Express, November 3rd 1968.

Chapter Four
When Face to Face?

Evidence with the truest ring has been when an incident has happened unexpectedly and unsought. Even if there could be some error of judgment in a detail or so. This possibility of a mistake in part of what a startled witness saw in less than a split second is probably the reason why serious investigators are divided as to kill on sight so as to obtain identification, or not to shoot. As referred to previously the "to kill" protagonists say that only getting a body for dissection will reveal its nature, unacceptable as this is to some, including this writer. Yet even among those who want to see a body for complete knowledge, one finds that they weakened at times when challenged.

One new researcher I met as recently as 1979, told me how in talk with a British Columbia hunter who had been advocating the killing theory, this hunter was not hesitant as to what he really would do if confronted with one of the species. I now had doubt whether he would shoot. He is one of those I know, and having noticed his milder approach to the whole problem when we last met, I think that these days he would react as others have done in the past; the very few who unexpectedly came upon a Bigfoot. They were mostly hunters in very desolate areas, and they said afterwards that it was impossible to fire, because in the quick infinitesimal glimpse they obtained, the creature looked so human!

A Sasquatch/Bigfoot sighting of a controversial nature occurred in 1974 at Harrison Lake head. I saw the Lake from the high mountains above it two years later when colleagues and I drove up the loggers' tracks to stop at the scene of a more recent sighting, which will be described after the 1974 account. Harrison Hot Springs village is on terrain surrounded by these mountains that are more inland than the coastal ranges. The Lillooet range above Harrison is North of the Cascades Mountains which start in the United States but loop up somewhat into Canada's North-West.

Harrison Lake is 44 miles long and possesses several forested islands which are seldom visited, some say, never. According to John Green[1], Bigfoot expert and author, and collector of relevant casts and relics, the Lake islands are teeming with animals and birds, some of which might be still unknown to man. Cargo vessels plied up and down the Lake to go ashore at the head, but had never landed anyone

---

[1] https://en.wikipedia.org/wiki/John_Willison_Green

on the islands, said Green. There may be alterations now, but even three years ago, had I wanted to take a look at the wild wooded area at the Lake's head, the project would have involved a helicopter trip which I could not afford.

And the wooded area at the top of Harrison's vast expanse of water was what interested me. But I had to merely look down upon it from the desolate heights where we had pulled up.

In 1974, a Youth Summer Camp was on the edge of a tangle of rough woodland and brush at the far lake end. The youngsters enjoying the rough surroundings were in a few cases slightly disturbed patients from psychiatric care, sent there to help their condition. By and large, they were a happy group, and their leader, Wayne Jones, ran the camp and staff successfully.

One of the girls, fifteen-year-old Claire Nicol, was returning to the group of huts from the beach, when she encountered a huge hairy creature crossing the road only fifty feet in front of her. It was covered with black hair, was very wet, and seemed to have come shambling up from the creek, and she told in great panic how two of the boys with her had promptly turned tail and left her.

Some of the camp staff, including Wayne, said they too saw it before it vanished again in the brush behind the camp which was right on the forest edge. It was reported to be seven or eight feet tall with long arms and thick dark hair- not fur, some said. Smoke and glare from a campfire had caused this sudden emergence from the thick wilderness behind the club houses.

The youngsters got so excited that their leader warned them not to talk about it anymore, and investigators who came to the scene had only contradictory accounts, and eventually complete denial of anything unusual having happened. There were certainly exaggerations in the young people's stories, and the leader shut down on it completely and would not speak to anyone who tried to unravel what really had occurred.

The whole report was put down to the fact that the young people being a little sub-normal, had probably blown up some slight unexplained incident into something that built up to a Sasquatch story. So, this incident was reduced to being a non-event, like others in the past.

The Lake Harrison youth summer camp was shut not long after, though when I was in British Columbia again in 1978, I heard it had been re opened.

For months, talking to different interested persons and people familiar with the area, I could not be sure in my mind of what it was that had caused a panic at the Harrison Lake camp. An invention of excitable, sub-normal youth? What had started it off? No. It did happen.

Very different is the story of Merlin Hellener. It happened at a summer youth camp too, but a very normal one in the Cascade Mountains, in the summer of 1962 or 1963, when she was about eleven years old.

She wrote to me comparatively recently after reading some of my writings about the Bigfoot question. She is now a professional woman living in California. She told me that she was intensely interested because of her experience in that Cascade Mountains camp when a child. She now believed that she had seen a Bigfoot and told me about it.

During her first few days at the camp, she overheard some adults talking about food disappearing from the kitchen, but paid little attention at the time.

The camp was laid out with chapel, dining hall, and kitchen in a broad clearing. The girls' cabins were staggered in a cluster back in the woods with bathrooms between them and the scattering of boys' cabins which were also actually inside the woods, at a forest's edge. The entire camp population - boys, girls, and adults - would frequently remain in the vast dining hall after dinner, singing and playing games. On one such night when the entire camp was in the dining hall, Merlin left with two of her friends to collect some Kleenex one of the girls had left in their cabin.

The cabin had bunk beds with sleeping bags on them arranged along either side wall and along the back wall. A large, hinged window along the back wall was always kept propped open with a stick.

The three girls walked into the cabin in darkness, opened the door, switched on the light, and saw what Merlin Hellener now believes was a Bigfoot. He was crouched on the top bunk in front of the window. She still remembers what seemed like a suspension of time as the girls watched him slowly, so slowly, turn his head to regard them with an expression which was not malevolent, nor frightened - but – deeply penetrating, intelligent, and curious. He seemed huge nearly filling up the bunk, dark and hairy, powerfully built, and was (this image, she said had always caught in her mind) on his hands and knees. The girls all began to scream at once and ran in panic back to the dining hall to gasp out their story. They returned with the staff in great excitement to find the cabin quite empty. But there were leaves, twigs, and dusty prints on the sleeping bag, the stick in the window had been knocked down, and underneath the window outside there were prints in the soft dusty earth.

Merlin Hellener could not remember after the passage of years if the prints were first discovered by flashlight that very night, or if they were found the next morning, or if it was Merlin herself who found them the next day.

She told me she did follow the footprints the next day for some way into the woods until she got frightened that the creature was watching, and she ran back to the camp. She also, in good detective style, to which as she said in her early days, she was very keen, did several drawings of the imprints to take home with her. When she got home, she compared her drawings with those of her brother's Boy Scout Handbook. The footprints were of two sizes, oval-shaped with round toe-marks around them. She did not find any heel marks, so to her they did not look like human footprints. There were definitely no claw marks. What they most resembled were the marks human feet make when someone is walking on tiptoe.

The adults at the youth camp had at first suggested that the children's imaginations were working overtime, but since it was obvious they had seen something, they kept telling them that they must have seen a bear. Now, Merlin grew up in mountain country and had seen bears at close range, and certainly knew that this was not what they had seen. All three girls agreed that it looked like a gorilla, but when Merlin insisted that it must have been a gorilla since there was a similarity, the pastor's wife took the child aside and threatened to send her home if she did not stop upsetting the other children!

On the night of the incident, the camp was full of disturbances - bats, porcupines, rats, kept popping up in one cabin after another, said Merlin. Lights kept getting switched on, and no one got much sleep. Whether or not all this related to their Bigfoot, she did not know.

The next day all the children were told the woods were now absolutely out of bounds unless they went on a camp-sponsored walk with adults, and any children caught beyond the boundaries would be sent home. The rest of the time was spent without incident. But for several weeks after she came home Merlin was afraid to be left alone in a room because she had got it in her head that Bigfoot had followed her home and would come to get her. Her brother went to the camp a week or so after she had been there, and he told his sister afterwards that he overheard some adults talking about something being wrong about the camp, and that they were thinking of shutting it down.

At the time she had never heard of Bigfoot, and the incident remained half forgotten for some years. Then when sixteen she happened to come across a reference to the phenomenon, and then even later she read the writings which prompted her to send me a letter in which she told me of the experience in the Cascades.

"In retrospect," she told me, "I could imagine three separate explanations for what we saw. It could have been a human dressed up in a gorilla suit. If so (far-fetched as it would be) it does not seem practical to me that he would run away on all four feet instead of two. Secondly it could have been a gorilla although gorillas are not indigenous to the Oregon mountains, and it does not seem likely that one could have been lost from a zoo or a circus without the fact being publicised. It certainly did look real, whatever it was, and had intelligence in its face."

Merlin ended this communication with the comment of how often down the years she had thought how fascinating an experiment it would be to locate the two girls who shared the experience and compare memories. But how locate them? She would not know where or how to begin.

And this report from Merlin Hellner made me think again of the young people's panic at Harrison Lake summer camp several years later.

And in more recent years in 1976 and 1978, I wondered again about the Harrison Lake story as I looked down on it a few thousand feet above its moonstone-blue waters.

We got out of the station wagon, John Green, Brian Leach, photographer-journalist and I, and took pictures on the spot where that other comparatively recent encounter with Sasquatch had allegedly occurred.

A couple in a small car had ventured up the loggers track some months earlier. While the man got out, leaving his companion in the car, a huge hairy and very dark creature had come bounding down through the heavily forested peaks on the left of the car, flashed past the windscreen, and went crashing down into the thick undergrowth of the deep ravine overlooking the lake. She screamed out that she had seen a Sasquatch, and her husband hurried back.

Just where we were now standing.

She had glimpsed the creature for a fraction of time. She was the only witness. This is always open to query. Had she mistaken something else for the fabulous creature that has puzzled the centuries?

As happened with the Harrison Lake youth camp, the story was questioned. Then the couple went back to Seattle, their hometown, and refused to speak. The fear of ridicule and censure is still very great.

I took a last look at far-away Harrison Lake head where the youth camp children had been silenced and told not to gossip and make up stories. Comparing this discounted sighting with Merlin Hellener's story, I think those sub-normal youngsters did see something that scared them. In their mental state they would not have had the guile and brain power to make up a horror story, as has been impressed upon investigators by the camp persons responsible. I think they decided to deny anything abnormal.

Sasquatch/Bigfoot and equivalent species known by their many other names are reported to be swimmers. They have been noted to emerge from deep caves or cliffsides on the wilder reaches of the Pacific west coast to enter the sea to catch fish. Glimpses of them have occurred on riverbanks too, as some United States reports have shown.

John Green referred to one of these theories. He told me he thought that the Sasquatch and its equivalent elsewhere, is not a possible relic of perhaps very early Man but is entirely animal. While true Man can think and progress, the Sasquatch or Yeti has no brain power.

Yet here are curious contradictions.

We discussed the Albert Ostman story of that prospector of many years ago having been captured by a family of ugly primitive mountain giants. They had had motives for their kidnapping. They obviously could apply some reasoning for their actions. Yet in spite of his "animal" theory, Green said he considered the Ostman story to have been a true adventure.

Double opinions are often difficult to integrate logically.

Albert Ostman's extraordinary experiences occurred very long ago, and he only spoke of them in the 1950's when they were certainly distorted and probably exaggerated by the more frivolous elements of the media.

The Yeti phenomenon cannot be completely animal. I thought it was once. Now with added knowledge in mind I think it is a more complex phenomenon.

The late Professor Boris Porshnev used to state that a puzzling feature in Yeti analyses was in the conflicting characteristics. These showed features recognised as belonging to both man and beast. Even all those years ago Porshnev was saying what present-day researchers are beginning to realise.

The reluctance in official quarters to setting up world-wide enquiries are clear and convincing. The first obstacle is the great expense such a survey would involve. The second is political resistance. Government elections. One needs an independent millionaire public benefactor to science's Anthropo-Zoology to come forward and pay the bill that an independent world-wide expeditions-slanted survey would cost.

Meanwhile the Yeti image remains, though, many quarters would like it buried and forgotten. My S-Map, the outlines I discovered and designed some years ago, covering sightings, sounds and reports, still holds good. The theory and occurrences cover from the Caucasus to Mongolia, a fringe of the Tibet-China borders, Siberia, and the Himalaya. Then over the Bering Straits through Alaska and down to British Columbia and Northern California.

During the 70's renewed Bigfoot interest blew across some remote parts of the United States, and across some not even very remote.

One case that comes to mind arose while I was on that side of the Atlantic before I visited British Columbia. This was in July 1973 when a very odd monster-sighting report came out of Maryland. An editor I know who wanted to talk to me about it just as it had happened tracked me in my wanderings near the Canadian American borders as persistently as if I were the monster in question.

The creature had appeared to local people near dense forest outside of Baltimore, and someone had immediately given it the peculiar title of "Snally-Gaster"[2]. There seemed such a joke element in this probably folk derivation that I at once thought this incident was only what Dr. John Napier[3] in his book 'BIGFOOT'[4] rather

---

[2] RF - The Snallygaster is a folkloric creature with bat's wings, a long beak filled with teeth and a single eye in its forehead. It makes a noise like a train's whistle. The name is derived from Schneller Geist, German for "quick ghost.". German immigrants settled in the Frederick County area of Maryland in the 1730s. It has no connection with sasquatch.
[3] https://en.wikipedia.org/wiki/John_R._Napier
[4] https://www.amazon.co.uk/Bigfoot-Dr-John-Napier/dp/0425033813

engagingly had described as "goblin territory" whenever he alluded to fantastically incredible reports.

But this one apparently was not. The forest confrontation was described more objectively by an observer on the spot who was a member of one of those occasional amateur Bigfoot-sleuthing groups. This man, John Lutz, said: "We know that the creature was seven or eight feet tall, and must have weighed between 350 and 450 pounds. It had self-glowing eyes". This seems to be a peculiarity described about several encounters after dark. This is an animal characteristic. The mystery creature appeared near a place called Sykesville. "It walked upright like a man", Lutz went on, "but was covered with black hair and looked like an ape. It smelled of decomposed flesh. If anyone wonders how we can be so sure, I and a consultant and three of my associates saw the thing standing at the edge of a heavily wooded area at 10 p.m.

Much of John Lutz's information I obtained personally. "We were searching near a Military Base," he told me later on, "And two of the officers were with me. They stated afterwards that they did not think their weapons would have stopped the creature had it charged us. While we were staring at it, the thing, whatever it was, turned and raced away into the woods. We tried to track it with dogs, but the dogs refused to go after it.

"The Military thought that it was some kind of gorilla escaped from somewhere, but we could not find any such escape had been notified anywhere. This creature is not a local legend. There has never been such a thing reported in this area. We are almost convinced that this is a stray 'Sasquatch'. We have picked up similar reports in Pennsylvania Virginia, and West Virginia in the last five years, and we are attempting to track down these reports."

The editor who had put me on to this story did not print two lines about it, but it was seized upon the same day by one of the late-night television programmes in Toronto. I had then returned to my hosts' home in Toronto. The TV presenters and others gave us some fun over the mystery apparition's weird nickname but did not explain its nature.

A few months after the Maryland story came a similar one from Illinois. It became known in November 1973 though the actual occurrences were some weeks earlier. The area was in Murphysboro' town locality. Series of related incidents were reported in New York. Other regions picked them up for the media as far as the recognised Sasquatch venue, British Columbia.

Mrs. Nedra Green was going to bed in her remote farmhouse near Murphysboro when from an outside shed came a piercing cry. Mrs. Green is said to have just remarked, "There it is again." People in outlying rather deserted country are accustomed to hearing unspecified sounds. At another home in the district, four-year-old Christian Baril was chasing fireflies in his father's backyard. He seemed to be out rather late for such a youngster. He called out. "Daddy, Daddy, there's a big ghost out back!" Christian certainly saw something. Other people around experienced mystery confrontations. Randy Creath and Cheryl Ray were a young couple standing in her darkened porch, when something stirred in a nearby brush. Cheryl turned on a light, and they both stepped back to see what had interrupted their cosy get-together.

A shape charged forward from the bushes. The two wide-eyed teenagers were face to face with an eight-feet tall creature with long shaggy matted hair all over of a dirty white colour. It smelled of river mud. All three stared at one another, the man and girl very frightened. Suddenly the creature turned away quite slowly, and then went crashing through the brush towards the neighbouring waterway known as the Big Muddy River.

For weeks before this particular sighting, Murphysboro monster had baffled police and scared residents. The first scare had begun before the Randy and Cheryl encounter. This had been just after midnight on June 25th.

A young couple were sitting in a parked car near the river's boat ramp. They heard a loud cry coming from the woods. Local people, hearing it too, had said it sounded like an eagle's shriek, which does seem rather unnatural, but perhaps in some circumstances eagles do give voice, or the American variety differs from Scottish, European, or African species.

What really mattered was what the couple in the car suddenly saw. They were shocked at the sight of a light-coloured, hairy, seven-feet shape lumbering towards them covered in river mud. They did not stay, but drove off hastily, and reported the encounter.

Later, when interested inhabitants wanted to hear more from them about the muddy, animal-like giant, the police diverted enquiries, saying the couple had "left the area". Their names were not quoted so as to save them embarrassment. The girl was married, but not to her nocturnal companion.

When the police and other officials inspected the area around forest and river, they found unusual footprints in the receding river mud. Officer Jim Nash was convinced that there was an unidentified monster roaming Big muddy River's wilder shores and woodlands.

It was the night after the Muddy River incident that little Christian Baril saw his "big ghost" and that Cheryl Ray and Randy Creath had their confrontation outside Cheryl's porch. Later on, Randy even drew a rough sketch of the creature.

"I know it's out there somewhere," he said. "It would be fascinating to study it again. but I hope it - he - doesn't come back. With everyone running around with guns and sticks he wouldn't have much chance, would he?"

Which shows that there are quite a few conservationists around, even in regions where a gun has generally been the answer to anything unexplained which might attack.

Toby Berger, the area's chief Police Chief, finally ordered his entire 14-man force to mount a night-long search. Jerry Nellis, a dog trainer, brought his 86-pound German shepherd hound, Reb. The hound found gobs of black slime after, with searching floodlights they had discovered a rough trail in the bush. This was on a direct line between the river and Cheryl Ray's house. Quickly, the dog was off down a hill, towards a pond and into the woods to an old empty farmhouse. The men followed and came to an abandoned farm.

At the door, the dog yelped and backed back in panic. Nellis threw the animal inside. Reb crawled out whining, and the men radio-ed for help.

Fourteen area police cars arrived, but the barn was found to be empty.

Police Chief Berger says: "Many things are unexplained, and this is one of them. We don't know what the creature is, but we believe that what all these people saw is real. We have tracked it. The dog got a definite scent. The people round here are good, honest people, and not liars. They are seeing something."

He discounted the hoax suggestion which nearly always comes up in such unexplainable occurrences. It could not be a trick, for the good reason that Murphysboro is hunting country, and nobody, however prone to practical joking would pretend to be a monster and risk getting shot. "And who would walk through sewage mud for a joke?" was his final comment.

In the history of the Yeti, or Abominable Snowman in more than one land and harking back to past scattered reports, there have been accidental killings. But such accidents involved the quick destruction of a one living body of bizarre unidentifiable nature, and in remote parts of the earth. So, nothing has remained as far as knowledge goes for study and conclusion. To the primitive and ignorant, hurry to remove all traces was no doubt to avoid awkward questions. For some others, to save embarrassment, while some might have feared censure because they were not sure if they had destroyed a human freak or some rare unrecorded animal.

And the hoax theory is probably decreasing. A hoax would not continue for so many years. When the vastness of both Americas is considered, the complicated machinery required to mount a nationwide consistent trick would be unthinkable, especially when such an exercise would bring no financial return.

Invented monsters are too expensive, and so the unsolved question has remained. And serious research is often hampered by local indifference. When communities in remote areas see something unusual that keeps recurring sporadically, they grow accustomed to it, and it becomes a habit. Enquirers from the outside world seem to make a fuss about nothing important.

In the distant past in the wildest regions of Russia, in densely growing forest, mountain ranges and in deserts, the Almasty were considered entirely animal in spite of their half-human appearance. And they would be hunted and killed by some tribes who were even more savage than they were.

Chapter Five
More Encounters

The swamps and beyond of the Everglades of Florida are better known generally than that Big Muddy River mystery on which the last chapter closed. But swampy and muddy were some of the local experiences of a few past years. Though the outside world heard little about Florida's own type of Hairy Man, and if they heard rumours, they put them down to Indian folklore which is certainly abundant but does not explain all local incidents.

H. C. Osbon was prominent once in an organisation called The Peninsular Archaeological Society, and he described an incident that took place in the Big Cypress Swamp. He and fellow members were digging for Indian relics. They were not looking for anything else, but according to their account something came, and had a look at them.

This was a seven or eight-foot-tall manlike creature that smelled unpleasant, and was very hairy, the moonlight picking out contours clearly. The description tallies so much with descriptions of previous meetings with phenomena as to become almost monotonous.

The phenomenon surveyed them, and then made off in the Swamp, leaving enormous tracks which they discovered next morning. They took plaster casts, as other amateur and professional anthropologists and zoologists keep doing in both Americas, and elsewhere too. The Russians need only be mentioned in passing here, as it is known that their interested parties have been on this exercise for years. They are mentioned in later pages.

Belonging to the Miami Museum of Science at this time gave Mr. Osbon a certain status to his and his fellow-members' findings. The Big Cypress Swamp story was kept quiet for some time so as not to cause alarm. It was finally released for the sake of public cooperation in case of similar incidents occurring again at or near the Swamp.

They did. Not long afterwards an unusual entity was being hunted by a dozen or so local people, for something large, unknown, and mysterious had scared children one evening while they played on nearby uncultivated land. There was a housing estate in this area, a few miles west of Fort Lauderdale.

Children had described the apparition as having an ape face, very long arms, grey patches, and they insisted "It was much taller than Daddy".

Nothing more happened, and following the familiar patterns, the scare died down.

Old Indians speak of a swamp haunting they called "The Skunk Ape" because of its offensive smell, but a large majority of people have always denied the tales of such current sightings as nonsense. Others might have had reasons to doubt this but kept their own counsel for fear of being laughed at.

But Mr. Osbon must have been impervious to local jokes. He was said to be a down-to-earth and practical businessman interested in electronic engineering as well as in his archaeological hobby and was not afraid of ridicule. After his Society's Swamp experience, he believed there was "something".

The Los Angeles Times in January 1974 commented on the scarcity of reports of big manlike creatures being spotted from time to time in isolated regions of the Pacific North-West. This had not diminished the interest of primatologists and lay searchers. For, in a few cases, the natural sciences were trying to identify the Bigfoot reports.

Dr Geoffrey Bourne[1], Director of the Yerkes Regional Primate Centre, was believed to have said that he thought it possible that seven and eight-feet humanoids such as Bigfoot or Sasquatch of the North Americas, and the Abominable Snowman or Yeti of Asia might really exist, roaming inaccessible regions remote from the paths of men. Dr. Bourne was one of the experts who in 1967 had viewed the Roger Patterson film of the Californian Female Bigfoot. Like many others, he was not quite convinced, but had reserved his judgement.

In spite of the Los Angeles newspaper's view that Bigfoot news were sparse, 1974 brought a number of scattered reports of encounters. Most came from Illinois. A disabled war veteran saw two of the creatures, and described one of them as grey and hairy, but shorter than previous sightings relate. The most remarkable, but often noted feature was that this Bigfoot had pink reflecting eyes. This peculiarity like a sort of phosphorescence, occurs in more than one report, from data by Linnaeus down to current records. It suggests a species that operates more during hours of darkness than in daylight. Carl Linnaeus's[2] work in this sphere is presented in a later section of this book.

---

[1] https://en.wikipedia.org/wiki/Geoffrey_H._Bourne
[2] https://en.wikipedia.org/wiki/Carl_Linnaeus

Yet encounters of the Yeti type do occur in daytime, which indicates that types no doubt vary, some functioning during hours of clarity while others, perhaps timid, follow their life pattern when darkness cloaks their movements.

Among the hundreds of eye-witness accounts John Green collected was one from William Roe about a daylight encounter several years ago. This experience stood out among many because Roe, a hunter in British Columbia's most remote areas, had been able to watch at close range a Sasquatch that did not know at first that a human being was present. Roe was also the first in these wild North-West regions to describe a Sasquatch as an ape-like, near-human being rather than as a giant Indian, which was a favoured image encouraged by their lore and myths. Green said that though this idea was liable to provoke laughter among the more informed of the non-Indians, it is possible that the Indians put out the Giant Indian legend to stop arguments with the white men.

Roe had his gun at the ready when he saw the Sasquatch, but after wards he said that he could never have shot it. It would have been like attacking human like himself.

Related to extraordinary meetings, one could now refer back to the strange experience of Albert Ostman in the Canadian North-West.

While I kept reports of this incredible and ancient story after it had happened, I never wrote about it. Transatlantic popular magazines had coloured it up to "goblin territory" and I fought shy of it.

It must have been in the early twenties that Ostman went on a one-man hunting and mine-discovering trail after working for several months on construction work. He took a guide familiar with the mountainous and deserted terrain he was searching. This guide was an old half-breed Indian who had foraged in the vicinity all his life. He knew part of the way up the mountain leading to an alleged deserted gold mine.

This one had been discovered by an eccentric prospector who found it and then vanished.

What happened to the prospector, Ostman asked his companion, who replied that he must have been killed by Sasquatch.

49

And what was Sasquatch? asked Ostman, as if this meant some hurricane or meteorological disaster. The old man replied that they were large wild people like giants, and they lived high up in part of the mountains where nobody went.

Like many intelligent, knowledgeable, people then and similar people now, Albert had never heard of the Sasquatch in spite of his nomadic job.

As the Indians were noted for their tribal legends, Ostman just shrugged this off. After he had paid the old man and arranged for him to rustle up some stores for him lower down the mountain for his return in a few days, Ostman went up the steep track alone to locate the gold mine as others had done before him. But they must have turned back at one point where he now did, as conditions were getting rougher and rougher and there seemed no sign of discovery. Albert climbed on doggedly up and up, and then gave up his battle against the heavily forested and now quite dangerous trail.

He was pausing to get his bearings and to decide the best way to retrace his climb when the decision was taken for him.

He had heard no sound of anything alive approaching but was suddenly set upon from behind by something that must have been watching him from the brush. He was gripped by gigantic hands, bundled into what felt like coarse prison sacking or woven branches, and was trundled farther up the mountain, bumped from side to side, and unable to make out his captor.

At last, they stopped, and he was released, dazed and bruised, but still all in one piece. He saw that the creature that had caught him was an incredibly hairy and ugly giant, animal-looking, yet not an animal. Albert had been dumped down in a sort of mountain fortress, a rough camp where three other similarly unattractive beings crowded around to look at him, while the lot communicated with one another in grunts and snorts.

Ostman had still managed to hang on to his fairly substantial rations in the sack, and as they pressed around him, examining his possessions, he realised that this was a family. Apart from the giant there was an old mother, and a boy and girl. They were as ugly and inhuman looking like their parents.

They did not hurt Albert, but he soon knew that they would not let him go. For days he lived, ate, and tried to converse with his stone-age hosts in their primitive lair. The food he had brought fascinated them, and the brother and sister were more friendly than the old couple. One and all showed their anxiety that he should remain with them.

Apart from his food store, some of which they liked, and some that did not find favour at all, his damaged clothing and what he did with it was a constant source of curiosity and merriment. Then they would all squabble among themselves. By their behaviour, Ostman knew that he was the cause. He would always be watched in case he tried to escape.

One day by a successful ruse, he did succeed in escaping from their camp. One report of his adventure told years ago apparently, how he tricked the youngsters in some rough-and-tumble game on the mountainside, and it was during the antics that he made his dash for freedom[3].

According to Ostman's account of the few days he was their unwilling guest, they seemed to him more human than animal in spite of their hairiness, ugliness, and huge proportions. But they were very much stone-age products. Some third-person narratives about him later blew up the story of his stay with the alleged Sasquatch family to romantic dimensions. They said the old man had captured him to provide a mate for his daughter.

Though I never quoted the Albert Ostman story in my previous Yeti books because of its unbelievable nature, I present it now as in changing current opinion it might not be so far-fetched after all.

Indian tribes of the remote North-West of the Americas would once tell how their ancestors knew that communities of giant hairy creatures did exist at one time in the wilder regions of mountains and forests. But that these had diminished with succeeding decades. Yet, currently, a few Indians still say that they are not entirely extinct but exist in extremely small groups of twos and threes.

The Sasquatch family that captured Albert Ostman could have been strayed descendants of those giants of the far North-West who for some unknown cause had wandered down into the lower mountain ranges of North-West territories.

From that desolate North-West in 1927 came a story about the capture of a solitary Indian hunter by one of those reported giants.

---

[3] RF - Ostman actually escaped by feeding the big male sasquatch a whole box of snuff and running off as he was freaked out at the taste.

The Indian finally escaped wearing only his underwear in spite of bitter weather. Probably if this is not an exaggerated story which it could be, the Sasquatch had stolen his outer clothing together with his equipment and rations. The Indian reached one of the outlying missions in a miracle of survival. Though not old his hair had turned white from the shock of his experience. His character had changed, and he would never speak of his adventure. A similar adventure is given in Russian records quoted in a book of mine, "The Yeti".

The Cascade Mountains and densely forested country of Oregon is one of the areas reputed to harbour Bigfoot. In recent years, Tim Dinsdale[4], British author and Loch Ness expert, was visiting friends who lived in a bungalow on the highway which had heavy forest land on either side. When Dinsdale and a companion were leaving to return they said they would walk back to the Dalles Research Station which was then run by Peter Byrne, well known for his interest and work in Bigfoot investigations. The centre since then moved to the Hood River region of Oregon. Dinsdale was a guest of Byrne's at the time. It was near night-time, and the bungalow hosts said that they never went out anywhere on foot at night, but always used their car, as they would be too frightened to walk in the area after nightfall.

In spite of sightings reported with increasing frequency in the Pacific North-West and in other terrains, the only photographic evidence available from relevant territories is the Female Bigfoot film, pictures of discovered footprints and their plaster casts. The measurements are generally fifteen inches long by five to six wide. But there are smaller tracks too. And larger ones.

Dr T.D. McKown, at one time professor of physical anthropology at the University of California, once had some interesting comments to make on Bigfoot and footprints. He had spoken first of the Patterson film. He told how the story of mysterious, giant-sized tracks is an old, old story. Such tracks have been reported since the beginning of recorded time. Records indicated that millions of huge footprints have been discovered though the ages. Reports came from Asia, Africa, and South and North America. Some had been found in Europe too. He said how at times there had also been reports of tiny footprints, supposedly made by little people.

In recent years too giant footprints have been found, quite apart from those attributed to Bigfoot. A paper was sent to me a few years ago relating adventures of two young students in the Canadian Caribou National Park. This was a remote terrain where visitors were warned to exercise care as it was said to be the stamping

---

[4] https://en.wikipedia.org/wiki/Tim_Dinsdale

ground of Grizzly Bear. The students had excavated in a huge deep cave in spite of warnings, and there on rock they found imbedded the marks of a giant footprint which was not that of a Grizzly. There is a similar phenomenon seen in Norway. A friend of Norse forbears described it to me as a huge human-looking foot impressed on a rock in mountainous countryside.

Dr. Theodore D. McKown's writings on stone-age discoveries in the Middle East are of great authority too.

On a lighter note, when in 1976 explorer and author Colonel John Blashford-Snell[5] took his family on a Christmas holiday in Southern Nepal he found some very extraordinary footprints and feet in a small tribal enclave of Tharu tribesmen in the Terai district. The footprints were much in evidence and had been well preserved, and even the local temple had drawings of them on the walls. Those tribesmen's feet were distorted, with enormous, big toes standing out at right angles. It seemed a few other people in Nepal had this trait. It is a defect, a Mendelian recession known as Hallux varus.

The place was near a small village called Mehrauli, several miles from the nearest town, Narayangarh. There is an airfield at Mehrauli, and just a few huts. The temple's guru, or holy man, possessed the same defective feet.

John Blashford-Snell told me how he spoke to the local inhabitants. Some were morose, others friendly. One or two of the huts had sketches of the local feet. It seemed quite a cult. One local character, pointing to the wall drawings, said: "Yeti" and he laughed.

Unbelievers in the Yeti phenomenon were quick to suggest that this freakish deviation was what had started all Yeti reports, which is irrelevant, as it cannot account for centuries of the Bigfoot mystery, this human defect existing in only a small part of India's and the world's vast continents.

I personally vouch for the above incident. Facetiousness does at times underly certain reports of sightings, and such reports form witnesses and officials are quoted for what they are.

---

[5] https://www.johnblashfordsnell.org.uk/

An American family saw what they took to be an erect hair-covered man standing at a thickly wooded area edge about 200 yards from their home near Friendsville. That day they watched for about half an hour. The next day being warm and sunny two of the children were playing near a field. They heard a deep growl and looked towards the border of the woods. Two hairy beings that could have been men or apes moved slowly towards them, and as the children screamed their father heard them and saw the hairy creatures chasing his children who were running back home. The father took down a gun which he kept in a barn and fired at the unidentifiable but very solid beings, but they fled back into the thick woods.

Mr. Miller described them as with black or dark hair all over their bodies, of six or seven feet in height. He said this was the most frightening thing he had ever witnessed. The creatures were never seen again, but they were heard emitting terrifying screams between midnight and 5am. The sounds came from dense woods which are part of the Alleghany Mountains near where the Millers had their home. They may live there still.

The "visitants" never returned from their mysterious forest hide outs. Though fields and woods were searched for clues, the only trace found was of a partly obliterated foot outline which was too poor to photograph.

The law officers of the County Sherriff's Department were very sceptical of the incident, just as many official sources have been over other peculiar occurrences which had happened earlier on in Louisiana and Missouri in July 1972.

Communities in heavily wooded areas kept reporting sporicidal confrontations of startling nature. These were invariably with hairy bi-peds of unpleasant odour, discordant voices, and "glowing orange eyes",

One incident was at a place called Marzolf Hill[6] where Terry Harrison, a little boy, saw "a big hairy thing" in the family's back yard. His father, Edgar Harrison found no trace of the "monster". Yet it had already been glimpsed by neighbours who soon were nick naming the creature as "Momo" - a derivation from Missouri Monster.

More sophisticated members of the community suggested this could be mischief on the part of local youngsters, the high summer's atmosphere being conducive to give people ideas when time hangs heavily. Edgar Harrison who had worked for 21 years

---

[6] https://missourilife.com/the-legend-of-mo-mo-the-missouri-bigfoot/

for local boards of public works, tried to discover what his son had seen. He failed though as he said: "I spent my noon hours with some of the fellas looking in the woods. I'll look under every piece of brush, every piece of rock, and won't stop until I find what this monster business is all about." But he never did.

Another encounter had a more down-to-earth sound about it and is in my files. Ellie Minor, an old, grizzled fisherman, was cleaning out fish at night in an isolated enclave of woods, hills and stream. His dog growled a warning, and Ellie looked up. He described what he saw. The "monster" was "standing there, black as coal, with hair down to its chest! I threw my torchlight to play upon him, and he turned and whirled off thataway, vanished in the pitch dark...." Ellie said how he tried next day to track the creature, but the ground was too dry.

Patrolman John Whitaker, an easy-going, good-tempered man, was patient and amused at the flock of newsmen and sightseers who hurried to different "monster" locations, and Ellie's vantage spot. But he said: "I've known Ellie Minor all my life, and I've never known him to make anything up. Something just might be in those hills..."

Something might.

Here again was this peculiarity of the inadmissible Missing Link's glowing luminosity in the eyes when confronted in darkness, whether the basic eye colour appears red, gold or green, as observed by more than one witness down the centuries.

Part animal, part human? The Relict Hominid of Anthropo-Zoology?

Chapter Six
A Scientist's Bigfoot Documentaries

Some years ago, Professor Grover S. Krantz sent me some of his papers based on many years of research in the field. He is of the Department of Anthropology, Washington State University, and attended and spoke at the Sasquatch conference at British Columbia's university in Vancouver in 1978 which I attended.

He is one of the few established anthropologists who have applied their anatomical knowledge and experience to serious practical study of the Bigfoot situations. In his own words:

"While a few scientists take the reports seriously, the majority doubt that such animals exist. Sightings are usually dismissed as being of standing bears, hallucinations, or fabrications. Footprints are explained as those of bears, or if they are quite distinct, as deliberately planted hoaxes. Sceptics point out that no specimen, living or dead, is available for scientific examination - not even a single bone has been identified as belong to this species."

Professor Krantz stresses that the only tangible evidence for this type of primate consists mainly of a few photographs and here one can insert the fact that these were unsatisfactory - the short movie strip of the Patterson Snowwoman, and some handprints. These last were taken by Krantz himself in 1971. He also referred to footprints which have been measured, sketched, photographed, and cast in plaster in several instances.

The name, Sasquatch, is of Indian origin. Bigfoot, the American term, is an obvious choice. He said that the American characteristics of the unspecified primate are the same as those of the Canadian species.

He deduced that if the animal existed, it would probably be man's closest living relative, and a member of the zoological family, Hominidae. It is not human, he stressed, adding that all reports were unanimous on the creature's lack of language, artefacts, and social organising.

How are we sure?

He considered that photographs of footprints which were examined in detail proved not to be simply enlarged human tracks, but showed various peculiarities, including flat arches and enlarged heels. The body weight of whatever made the tracks would affect the impression they left. Krantz described how within the last century hundreds of people reported seeing large, manlike animals in the forests and mountains of the American Continent's Pacific North-West. Hundreds too have seen the huge footprints left by these hairy creatures. It is possible that ten times as many people have seen these giant primates and their tracks without reporting anything to newsmen or law officers for fear of ridicule.

The typical Sasquatch or Bigfoot footprint, he said, looks very much like a man's, increased to about seventeen inches in length. Closer examinations show other differences. This has aroused curiosity, but there speculation ends. There has been little explanation.

A clue to these differences could be obtained by working out how body weight could affect a very large, but otherwise normal human foot. One distinction in the giant footprints is that they are quite flat. There is no instep, to speak of, which means the foot has no arch. The Professor drew attention that while adult Sasquatch footprints are flat, smaller tracks of supposedly immature species do show some arching. It is possible that the feet flatten with maturity because of the excessive weight of a full-grown creature's huge body. Professor Krantz, though, did wonder if there could be another reason for this flattening of the sole of the foot as the creature develops. Possibly a genetic mechanism which would lean to flat feet. But for what reason?

Personally, it seems to me the flattening process through increased body weight is the more logical cause.

Krantz made a special point concerning the many people who still believed the footprints were faked by hoaxers. Normally, fakers would not think of the flat foot peculiarity and would design indications of human arches to their artefacts for making the false footprints. None of the evidential footprints Professor Krantz examined showed a normal human arch, while showing oddities that never occur in the human foot.

Some Sasquatch prints show what is called a double ball at the base of the big toes. And in all footprint photographs, whether they were taken in the Himalaya or in the Americas, there is evidence also of an abnormally big heel. Many Sasquatch and Snowman tracks also show nearly equal-set toes, aligned straight across the foot in a non-human manner. Nonetheless, there are quite a number of Bigfoot tracks which do not indicate such traits but look more human. There may be a great range of sizes and toe-aligning from an almost human appearance to the primitive, near-grotesque and unfinished look of typical Snowman foot tracks. The length and breadth differ too at times, from the first recognised 12 ½ by 6 ½ inches described by Eric Shipton to the sixteen to seventeen-inch dimensions relating to the Canadian and American species. Variation in size and type is just as possible in their case as in that of the human race.

Toes partly in a straight line can occur with humans, that is, when the big toe, second, and partly the third are aligned. This has been quoted sometimes by artists and chiropodists as the perfectly natural foot, but it is no asset to its possessor doomed to suffer from the often-unnatural footwear of civilisation. This is only my opinion.

In his papers, Professor Krantz said he believed there may have been a few footprint hoaxers, and people who have fabricated stories about the bi-pedal forest primate, but he thought that the majority of tracks and sightings reports were not produced by practical jokers. Laying hundreds of tracks in remote areas hoping that just a few might be discovered would be too unrewarding in sensation value, and too expensive.

Some of the media have reported Professor Krantz as saying that henceforth he would carry a gun and shoot at a sighting so as to acquire a body for investigation. I think this was exaggerated, and if faced with a human-looking creature, the Professor, like others before him, would not shoot.

Sporadic discoveries sometimes shed a pucklike humour on established beliefs. There could have been practical jokes even in antiquity. And the course of evolution is comparable at some points with the same process in the animal kingdom. In animal evolution diversities occur even in the history of one type of creature. These changes happen through environmental conditions such as migration to new habitat, forces of nature that commanded change or dispersal for the sake of survival. This in turn brings about variations in natural colouring and feeding habits sometimes.

This adaptation to surroundings for survival is shown too in Homo Sapiens progression.

When sceptical investigators in the Yeti question propose explanations, they offer various types of fauna as candidates. The bear has always been the most popular choice, for though this animal normally progresses on all fours it is bipedal at times. So, it is possible that some of the Abominable Snowman sightings were occasionally bears seen from a distance and standing on hind legs for a short moment. But it is not a practical solution every time there is a Yeti, Bigfoot, or Sasquatch report. Other candidates were the Langur Monkey, even the fox, and the Dremo. This is the picturesque name in some regions for the rare Tibetan blue bear, figuring in some ancient Indian legends as the kidnapper of young maidens! It is actually a rather beautiful animal with a finer coat than that of the ordinary bear. There were only four skins in Britain a few years ago and I was privileged to see them. Blue as a name is somewhat a misnomer in spite of a slightly smoky tinge. The skin is more remarkable for a wide strip of golden coloured fur down the spine, from shoulder height to hind quarters.

Sometimes a human candidate appears. This is generally a genuine, long-distance mistake. The problematic figure was probably some wandering half-wit, or a withdrawn, wild-haired and ragged mountain and forest drop-out, half naked, hirsute, and preaching immortal truths to nature. Such eccentricities do happen, and remote communities accept them tolerantly as harmless freaks and self-exiles.

Nature can show different characteristics even in one species. One report of Yeti sighting could refer to its reddish hair or fur and squat height, while another witness in a different terrain saw a lumbering giant, blackly furred and shaggier. Science once ruled that race in man and animal descended in a straight line without deviation from the archetype. Until Darwin's theories, scientists had relied on classification rather than to giving a step-by-step attention to evolution. Classification is not enough, for evolution is what it says, it evolved, altered, however long drawn and slow the process may seem to the average concept.

It was after 1859 when the theory of evolution in its new concept was accepted that the co-relative theory of the Missing Link became popular. But rather like the later Abominable Snowman question, the Missing Link topic was often seized upon for flippancy and irresponsible surmise outside science. Even P.T.Barnum[1], the showman, produced various groups of suggested animals to represent the idea for public entertainment.

A few of the ideas were popular absurdities, even if they might have inspired the spate of searches for living creatures that might have filled the Missing Link gap and help complete the knowledge of what some quarters called the Chain of Being. Some little-known aborigines were suggested candidates, and other types of remote

---

[1] https://en.wikipedia.org/wiki/P._T._Barnum

living beings were presented to fill the gap between anthropoid and man. It could be asked here whether the Yeti under another name figured at all in this list. Perhaps he/it was in that identity parade but got lost in the jostling for a place.

There have been errors of judgment down the ages, coming from both the knowledgeable and the not so informed. Some animal species occasionally throw up a "sport", a specimen of its kind with unusual features, or of unexpectedly large dimensions. Such a creature seen briefly in difficult sighting conditions could honestly be described afterwards in distorted terms. This can be even more so if the witness happens to be influenced by superstition, tribal fears, or memory of some engrained racial legend.

Until 1969 the differences between man and ape were not yet functioning in relation to laboratory work.

A distinguished list of men next classified humans and other primates more correctly. Johann Federich Blumenbach[2] said in 1775 that "The innumerable varieties of mankind run into each other by insensible degrees ".

This flexible nature applies to all life.

The debated theory of a vital missing species in life's Chain of Evolution did not help, and science of the 20th century continued to question who or what the mystery species could be.

Attempted elucidation of the Yeti question of these days have been as varied and conflicting as the Missing Link controversies of the past, and even scattered reports at times sounded like legends, such as the famous Jacko tale of Yale, British Columbia, when in 1884, this infant so-called ape that had human characteristics was found stranded on an escarpment overlooking a railroad extension and was adopted by railway workers. How Jacko, who later died, though well cared for, ever got there was a mystery, but however manlike the young creature seemed, he must definitely have been ape, and not a young, strayed human covered with the typical hair associated with Snowman facts and legends[3].

Friends of mine a few years ago were camping in British Columbia in an area where new communications were being developed. The terrain was similar to the vast area

[2] https://en.wikipedia.org/wiki/Johann_Friedrich_Blumenbach
[3] RF - Jacko is now almost universally dismissed as a newspaper hoax.

of the Jacko story. Actually, they were in the Rockies, and nowhere near the place of the Yale ape mystery, which has remained so because no relics of the odd creature were ever found, even if he had become quite a pet. My friends took this Rockies vacation every summer and became friendly with the men working on the new communications being developed. Timber prospectors, loggers, and other work parties were there. The area was still uninhabited and normally desolate, and was very beautiful, perhaps because of its impressive vast emptiness. My friends were not Sasquatch hunting, but just doing what they enjoyed each year they took this holiday from their city, Vancouver. The men on the site were talking about the country around when they described a recent curious incident.

One evening at sunset a large hairy shape appeared looking down at them from a high and very inaccessible wild outcrop of rock and brush. It seemed to them part ape, part man, and it watched them for quite a while. The men below had no means to do anything, they told my friends, but could only gape back in amazement. Then the large hairy shape. vanished in a thick belt of trees.

This could be a repetition of many sightings or experiences including one which occurred about one hundred years before, and which I found accidentally while researching on another subject.

I was looking up and down the columns of the Daily Universal Register on its news sheet dated January 4th, 1785[4]. This forerunner of the London "Times" gave space to an unusual story from the Continent. It reported: "There is lately arrived in France from America a wild man who was caught in the woods 200 miles back from the Lake of the Woods, by a party of Indians. They had seen him several times, but he was so swift of foot that they could by no means catch up with him; till one day having the good fortune to find him asleep, they seized and bound him. He is near seven feet high, covered with hair, but has little appearance of understanding, and is remarkably sullen and untraceable. When he was taken, half a bear was found near which he had but just killed".

I have quoted this ancient report as printed and could find no sequel as to the wild hairy man's ultimate fate. By reference to the bear the Indians noted, and reference to lakes and woods, this unfortunate half-animal being had evidently been captured in American forests and shipped to Europe as a curiosity, just as the good-natured rail workers were planning to take Jacko to Britain to show him at fairs, kind and good-natured as they seemed to have been to the controversial and never explained capture.

---

[4] https://en.wikipedia.org/wiki/The_Times

Ruth Benedict (1887 – 1948)
American anthropologist and folklorist.

"The purpose of anthropology is to make the world safe for human differences."

Dmitri Bayanov (1932-2020)
Russian Hominologist and Author of
'In the footsteps of Russian Snowman'.

Vladimir Bogoraz (1865 – 1936)
Russian revolutionary, writer and anthropologist especially known for his studies of the Chukchi people in Siberia.

Copyright Atlanta Journal-Constitution. Courtesy of Georgia State University.

Dr Geoffrey Bourne (1909 – 1988)
Australian-American anatomist and primatologist

Igor Burtsev

Keen tracker who has travelled half the world in search of the Yeti.

Peter Byrne

Author of "The Search for Bigfoot: Monster, Myth, or Man?"

67

MADAME BLAVATSKY.

Helena Petrovna Blavatsky (1831-1891)
Russian mystic and author.

Professor Carleton Coon (1904 – 1981)

Professor Coon was an American anthropologist. Coon's theories on race were widely disputed in his lifetime and are considered pseudoscientific in modern anthropology.

Pierre Teilhard de Chardin (1881 – 1955)

A French Jesuit priest, scientist, palaeontologist, theologian, philosopher and teacher. He was Darwinian in outlook and the author of several influential theological and philosophical books.

Odette Tchernine (1897 - 1992)
British author, cryptozoologist, novelist and journalist.

Odette Tchernine (1897 - 1992)
British author, cryptozoologist, novelist and journalist.

René Dahinden (1930 – 2001)
A Canadian Bigfoot (Sasquatch) researcher.

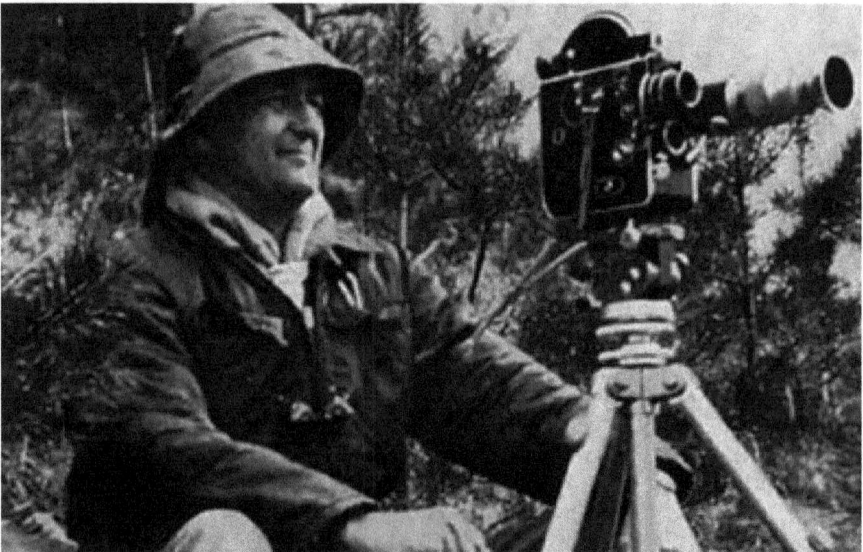

Tim Dinsdale (1924 – 1987)
A British cryptozoologist who attempted to prove the existence of the Loch Ness Monster.

Henry John Elwes (1846 – 1922)

A British botanist, entomologist, author, lepidopterist, collector and traveller who became renowned for collecting specimens of lilies during trips to the Himalaya and Korea.

Sir James Frazer (1854 – 1941)

A Scottish social anthropologist and folklorist influential in the early stages of the modern studies of mythology and comparative religion.

Bob Gimlin (born 1931)

Gimlin mostly avoided publicly discussing the subject from at least the early 1970s until about 2005 when he began giving interviews and appearing at Bigfoot conferences.

John Wilson Green (1927 – 2016)

A Canadian journalist and a leading researcher of the Bigfoot phenomenon.

Marjorie Halpin (1937 – 2000)

A U.S.-Canadian anthropologist best known for her work on Northwest Coast art and culture, especially the Tsimshian and Gitksan peoples.

Douglas Haston (1940 – 1977)

A Scottish mountaineer noted for his exploits in the British Isles, Alps, and the Himalayas.

Mirabehn (Madeleine Slade 1892-1982)
British born follower of Mahatma Ghandi

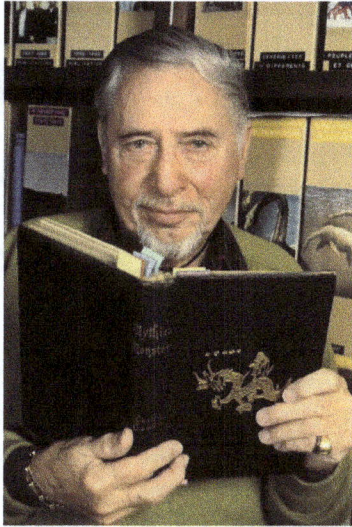

Professor Bernard Heuvelmans (1916 – 2001)

A Belgian-French scientist, explorer, researcher, and writer.

Joseph Dalton Hooker (1817 – 1911)

A British botanist and explorer in the 19th century. He was a founder of geographical botany and Charles Darwin's closest friend.

Dr Bernardo Alberto Houssay (1887 – 1971)

An Argentine physiologist who performed pioneering research on the endocrine hormones system or the "hormone feedback loops".

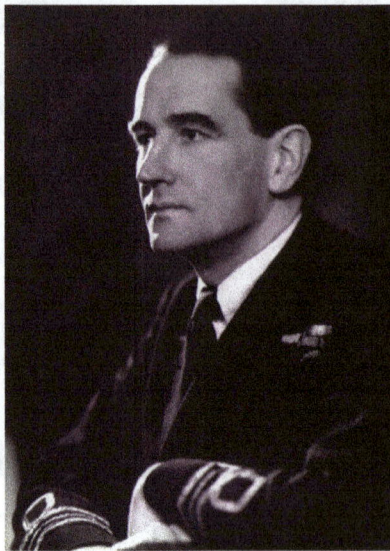

Ralph William Burdick Izzard (1910 – 1992)

An English journalist, author, adventurer and, during World War II, a British Naval Intelligence officer.

Major Kelvin Kent
A British adventurer, hiker, mountaineer, businessman and lecturer.

Marie-Jeanne Koffman (1919 – 2021)
A Soviet surgeon, mountaineer, and cryptozoologist.

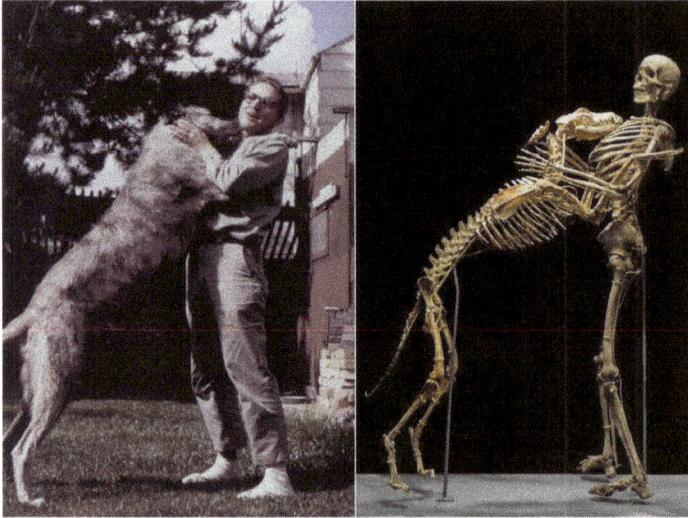

Grover Krantz (1931 – 2002)

An American anthropologist and cryptozoologist; he was one of few scientists not only to research Bigfoot, but also to express his belief in the animal's existence.

Claude Levi-Strauss (1908 – 2009)

A French anthropologist and ethnologist whose work was key in the development of the theories of structuralism and structural anthropology.

Margaret Mead (1901 – 1978)
An American cultural anthropologist.

"Always remember that you are absolutely unique. Just like everyone else."

Wilfred Noyce (1917 – 1962)

Cuthbert Wilfrid Francis Noyce was an English mountaineer and author. He was a member of the 1953 British Expedition that made the first ascent of Mount Everest.

Dr. William Charles. Osman Hill (1901 – 1975)

A British anatomist, primatologist, and a leading authority on primate anatomy during the 20th century.

Roger Patterson (1933 – 1972)

Patterson died of cancer in 1972 and "maintained right to the end that the creature on the film was real".

Boris Porshnev (1905 – 1972)

Porshnev took interest in cryptozoology and has been described with Marie-Jeanne Koffman as the "revered parents of Russian monster-hunting."

Colonel Nikolay Przhevalsky (1839 - 1888)

A Russian geographer of Polish descent (he was born in a Polish noble family), and a renowned explorer of Central and East Asia.

William Woodville Rockhill (1854 – 1914)

The first American to learn to speak Tibetan, and one of the West's leading experts on the modern political history of China.

Ivan Terence Sanderson (1911 – 1973)

Along with Belgian-French biologist Bernard Heuvelmans, Sanderson was a founding figure of cryptozoology. Sanderson authored material on paranormal subjects and wrote fiction under the pen name Terence Roberts.

Thomas Baker Slick Jr (1916 – 1962)

During the 1950s, Slick was an adventurer. He turned his attention to expeditions to investigate the Loch Ness Monster, the Yeti, Bigfoot and the Trinity Alps giant salamander.

Colonel Laurens Jan Van der Post (1906 - 1996)
A South African Afrikaner writer, farmer, soldier, educator, journalist, humanitarian, philosopher, explorer and conservationist.

Donald Desbrow Whillans (1933 – 1985)
An English rock climber and mountaineer.

Chapter Seven
Missing Links

Researchers can at times get bogged down in persistence to get at the truth about the Relict Hominid. Descriptions in certain cases indicate a pre-hominid creature rather than that first animal image of the Himalaya in the western mind. Yet in some areas of the Himalaya the image of the wild hairy man prevailed, while the later Professor Porshnev from his studies and experiences of a lifetime which proposed affinity with variations with different types of Yeti, but also admitted to the perplexing human characteristics some of the types showed.

This is why in current and future records, human as well as animal progression must figure side by side.

Varieties in the living world are flexible by nature. Flexibility applies to all life.

Some important scientific finds can have an indirect association with our subject.

In 1924 a skull was found in an African mine site, and was examined by Dr Raymond A. Dart,[1] who had just been appointed anatomist at Witwatersrand University in Johannesburg. After long and painstaking study, he made an exciting discovery. The Taung Baby skull[2], as it became known, showed distinct signs of early human characteristics. It had been discovered in the Taung semi-desert region of the Harts Valley area. The place where this happened had been inhabited in very ancient times by a community of baboons. Dr Dart considered that they must have been cave dwellers. The Taung site lay not 2000 miles away from other hominid finds. In a land of vast distances this is not considered a very great distance.

The discovery brought resistance from established experts and caused wide controversy.

The Taung infant skull showed marked differences from features characterising ape skulls. The bone structure and the positions of certain brain factors in this immature specimen indicated early human traits never found so far in any other ape skull.

Baby Taung's near-human qualities were eventually accepted, just as many years before the position of Neanderthal Man was grudgingly established.

---

[1] https://en.wikipedia.org/wiki/Raymond_Dart
[2] https://en.wikipedia.org/wiki/Taung_Child

Dr Dart called his early hominid discovery Australopithecus, and his history of Baby Taung and how he fought for its recognition, formed a later book about his long haul. It was published in 1959 under title of "Adventures with the Missing Link"[3] and written in collaboration with Dennis Craig.

The Taung child was definitely a missing link. Others followed. Notably the much later discoveries of the late Dr Louis Leakey[4] and of his wife in Olduvai Gorge, including the famous "Nutcracker" skull. Work which has been continued by his son, Dr Richard Leakey, also an anthropologist. At a research site he opened at Lake Rudolf in Kenya, he made new skull discoveries of very early origin.

The Nutcracker Skull (incidentally but most importantly eroded by Dr Louis Leakey's widow) could be termed as characterising Zinjanthropus, Paranthropus, or Australopithecus robustus, which are all three the same thing. Many other examples like this of the same fossil species are now known.

These acknowledged landmarks in man's long history have at times run parallel with "inadmissible" Yeti discoveries where similarities occur.

Here is an illustration, but quite a cogent one. A comparatively recent book by Sonia Cole is called "Leakey's Luck" dealing with the late Dr Louis Leakey and his work. She is an authority in the related science. She speaks of the 1959 Nutcracker Skull, which proved according to Dr Leakey that the cradle of mankind was Africa, and not Asia. But this is not my main point at this juncture. What impressed me was that she described how the 1959 skull had a peculiarity not often mentioned; it had a raised crest on top of the head. A repeatedly quoted feature in Yeti descriptions has been the appearance of a definite pointed ridge on top of the head.

This is what I mean by parallel landmarks.

The first time an African ape, the chimpanzee, was observed with some understanding, and then referred to in writing, was in 1699. The author was Edward Tyson, one of those travellers of all nations who are insatiably interested in the world they live in. He wrote a book of his observations in tropic lands entitled "The Anatomy of a Pygmie".

---

[3] "Adventures with the Missing Link" (Hamish Hamilton, 1959) - https://www.amazon.co.uk/Adventures-missing-link-Raymond-Arthur/dp/B0000CKFHO.
[4] https://en.wikipedia.org/wiki/Louis_Leakey

The great apes of Africa had often figured in returning sailors' tales, but they had generally been disbelieved. Tyson was more objective than the unbelievers at home. He wrote of his surprise that the chimpanzee closely resembled a man and differed more from other monkeys than from humans. This was revealed in one of his statements in which he said: "Our Pygmie is no Man, nor a Common Ape, but a sort of animal between both."

"Pygmie" was a simian term sometimes used by the early travellers.

Might not this early enquirer into Anthropo-Zoology have applied the same term to the mystifying, unacknowledged Yeti, even as Dr Dart's discovery and analysis of the Taung Baby Skull impressed him as being a Missing Link indication, though all Dr Dart claimed was that he had found a new family in anthropological terms.

Duration of the earth's eras run to millions of years each one, from the Pre-Cambrian of fire, steam, mist, and violent upheavals, and no life. Then the cooling down, seas, vegetation, first life, until aeons again went by, leading down to Man and his complex developments and predicaments. Natural selection ruled from the beginning, and at last came the Lower, Middle, and Upper Pleistocene. Man's world in primitive form but leading to Homo Sapiens as we are[*].

During the rolling on, merging, and altering life upon earth, is it not possible that in between all the mighty fragments of evolution some connecting life link could have been lost? And so created the mystery, some say, myth, of the Snowman, known of in opposite continents, called by many names, and uncatalogued. Where does the intermediate factor, the Yeti, slot in?

He must have been lost in more than one cradle of mankind, even of semi-mankind.

There were clues in the middle East, that central point joining several lands and forming a wide semi-circle enclosing the Mediterranean.

Well supported theories have claimed that Man arrived in Palestine, as it was known for centuries, before he got to Africa. Culture in pre-history established itself, travelling from east to West. Many Neanderthal finds occurred, and consistent excavations took place near the Sea of Galilee from 1925. There were important ones on Mount Carmel where Haifa now stands.

---

[*] Homo sapiens are not descended from Neanderthals, they are a sister group. The two species live side by side for many thousands of years and interbred. Modern non-African people have about 4% Neanderthal DNA.

The British School of Archaeology then situated in Jerusalem, undertook valuable excavation work. In 1931 what was described as a "prehistoric Necropolis" was excavated at Mugharet-es-akuo[5]. Skeletal remains were of several bones, an infant's skull and skeleton and a very thick skull of a fossil human. A male skeleton had the remains of a large pig near it. It might have been a ritual symbol to appease hunger. Or it represented some magical signal of food offering to sustain for the death journey - a frequent belief in immortality of a kind even among the pre-historic tribes of man.

There were western burials similar to this one, though without the pig symbol. In a cave at La Chapell-Sur-Saints, France[6], also in another French rock shelter at La Ferrassie.

Dr Theodore D McCown in Vol II of "The Stone Age of Mount Carmel"[7] told how acceleration of discoveries in what is now Israel brought new knowledge of the middle East Palaeolithic people of Palestine and the rest of the Near East for study in human evolution. This stage in humanity might have been crucial in mankind's biological progress.

There is still such divergence of opinion that even if the present day has a fairly accurate picture of ancient types of Man, much still remains in the dark. In a climate of uncertainty, much is taken for granted, and perhaps not enough room given for flexibility.

The Yeti mystery has run intermittently alongside accepted evolution. Because of its irregular appearances and lack of skeletal evidence of something unusual but possible, it has been left to the curious and adventurous to try to crack this alleged myth. At risk of being charged with repetition, where does this unproved Missing Link among the ages, documentations, and anthropological and zoological end of recognised men and animals, from the silent beginnings of our planet onwards?

Anthropologists have disagreed on the direct continuity of Neanderthal to Modern Man. Comparatively recent experts against the theory of direct descent were Boule

[5] Vol. One of "The Stone Age of Mount Carmel" by D.A.E. Garrod, & D.M.A. Bate - https://www.abebooks.co.uk/first-edition/Stone-Age-Mount-Carmel-Volume-Excavations/1083753220/bd.
[6] Boule, in "L'Homme Fossile de la Chapelle-sur-Saints. Anales de 1 Paleontologie 1911.
[7] https://www.cambridge.org/core/journals/proceedings-of-the-prehistoric-society/article/abs/stone-age-of-mount-carmel-volume-ii-the-fossil-human-remains-from-the-levalloisomousterian-bytheodore-d-mccown-andsirarthur-keith-pp390-pl-xxviii-text-figs-247-oxford-university-press-1939-price-3-guineas/6B058414181968E115B7D5758EA4F56D

in 1921[8]; Gieseler; in 1957[9]; Howells in 1968; Le Gros Clark in 1955[10]; and Steward and Valois. But Carleton Coon, the American Anthropologist, supported the theory of direct descent. Howell (no association with the previously quoted similar name) admitted the direct link, as did Hrdlicka[11], Pradel, Weidenreigh[12], Brace, and Weinert[13].

One of our suggested Relict Hominid clues may have fallen unrealised in the scientific gaps relating to the centuries.

I went to Israel partly because of an idea that the Middle East might provide one of the "parallel lines" of indirect clues to our subject. First came Galilee, a stay at the famous Scottish Hospice in Tiberias where in 1961 Japanese anthropologists had stayed. This was their base while they went out to research at Amud for a remnant of what was listed as Amud Man. They did find what they wanted and brought their silent discovery to the Hospice before taking it back to Tokyo University for scientific work.

The Japanese had descended upon Wadi Amud near Safed in a central part of the Galilee Mountains. The fossil remnants were found in situ in a burial grave. This was later described in "The Amud Man and his Cave Site"[14] edited by H. Suzuki and F.Takaie of Tokyo University. It was published in 1970 under the university's aegis and is prefaced by a delightful humorous poem by Dorothy Malvenan, one of the ladies in charge of the Hospice at the time the Japanese group stayed there during their dig. I read it myself there when the Hospice was managed by the Reverend John Murie and Mrs. Murie, very good hosts.

Amud Man fitted in a lower category than Modern Man, but he was of higher status than the Neanderthals of Europe. The skull's frontal bones formed a concave arch where Modern Man's is convex. He had a flat, pushed-back nose, and receding forehead. Some features equated with what the Yeti might have looked like, but the Yeti's oddity of the pointed skull was absent.

---

[8] https://www.amazon.co.uk/LAnthropologie-Vol-31-Classic-Reprint/dp/0259295434
[9] http://biodiversite.wallonie.be/fr/gieseler-w-1957-die-fossilgeschichte-des-menschen-in-die-evolutie-der-organismen-gustav-fischer-verlag-stuttgart.html?IDD=167775283&IDC=3046
[10] https://www.cambridge.org/core/journals/antiquity/article/abs/fossil-evidence-for-human-evolution-by-w-e-le-gros-clark-chicago-university-press-for-whom-cambridge-university-press-act-as-agents-1955-pp-181-20-text-figures-45s/57F8F16B02BC095C9D547F8C40128EC1
[11] https://en.wikipedia.org/wiki/Ale%C5%A1_Hrdli%C4%8Dka
[12] https://en.wikipedia.org/wiki/Franz_Weidenreich
[13] Vol. Two of "The Stone Age of Mount Carmel" byt T.D.McKown & Sir Arthur Keith, FRS.
[14] https://www.um.u-tokyo.ac.jp/UMUTopenlab/en/library/g_8.html

The Rockefeller Museum had a guardian whose memory was a treasure house of knowledge. He had met most of the great personalities in Anthropology and Archaeology who had come there. I do not know his name, but I must pay tribute to him for his cooperation in my search, for I was even allowed to stay there for quite a long time while I took notes from documents he brought out for my reading, and did my rough sketches which included the famous "Jericho Family of Fossils" discovered by Kathleen Kenyon and restored as in situ in one of the hall's show-cases.

The discoveries in Galilee brought evidence of Neanderthal presence outside Europe where the natural sciences there believed they had been confined.

Similarities in those Palestine species and their European cousins distributed comparatively far away may prove to some lines of thought that man's emergence was not in one place only, but was a scattered, multiple happening. This Mediterranean enclave of so many races' comings and goings may now be a repetition of the changes and upsurges that followed long after the Stone Age. Its geographical position makes it still the crossroads of a world where allegedly extinct ancestors had lived years ago.

And in these upsurges, destructions, and progressions, some skeletal remnants might have been overlooked or destroyed, and a clue to the Snowman might have been among them.

It is true that there are no accepted remains. My belief is that this if for the same reasons of predators and the forces of nature that operate. Animals, any corpse, is set upon and destroyed soon after death.

There could be one supposed proof of the Relict Hominid having existed. It came to light in recent years. It is the photograph of a female skull said to be that of an Almas excavated in 1953 in the Altai Mountains of Mongolia. As seen in some of the Calilee remains, this stone-age type is distinguished by certain similar features; the protruding top of the cranium, the heavier-than-man jaws. it may be discounted but was photographed and reconstructed.

Against it, or equating it, one could line up a certain Gibraltar skull, partly damaged, of a young woman of very long ago.

Before my Israel visit, I had been in Gibraltar. I took stock a little of the antiquity situation there, as well as paying the proverbial visit to the apes of the Rock. The skull in question had been found in 1848 in Ford's Quarry by workmen clearing a cliff site to improve the water situation, often a problem in Gibraltar. I was directed to Gibraltar's Museum, but found the skull was not there. On returning to Britain, I traced it to the British Museum's Natural History section in Cromwell Road, South Kensington, and was allowed behind the scenes where I sketched the specimen. Her infant's skull (Obviously her child) had been found next to her but was too fragmented for illustration. It had one clue to relationship with the Relict Hominid. The palate and back teeth were in good condition, and these teeth were so large and abnormal, that they reminded me of the huge teeth and jaws of the Almas girl described by Jeanne Koffman while she and her team were on field work in the Caucasus a few years ago.

Based on the discovery of those unusually large baby teeth, there might have been a whole pre-historic family on that much-contested gateway to North Africa ages before the apes took possession, or before every Mediterranean race landed and found the place good. Ages before the Virgin Mary, Our Lady of Europe, was enshrined as Gibraltar's Patron Saint on Europa Point, the southernmost cape on the Continent.

The Poet A.E. Housman, wrote the line "The tree of man was never quiet, Then 'twas the Roman, now 'tis I".[15]

This ran through my mind as in an unquiet age, I sketched the rather appealing broken skull, and thought of the baby with the giant teeth found near its mother, the first and forgotten Lady of the Rock - perhaps?

---

[15] From the poem, "On Wenloke Edge" by A. E. Housman.

Chapter Eight
Backwards and Forward

It is not entirely possible to keep a work of this nature tidily chronological. Pursuit progresses through new information, regresses to re-examine data, and progresses again. Mountains of wild terrain, animals, and sparse communities make unpredictable ecology at times, and are fickle taskmasters. More information on old records comes to light. Incidents and situations thought to have been dismissed or closed as unreliable, rear up again under changed circumstances, and inconsiderately produce sequels that cannot be ignored.

In 1974 Peter Byrne sent out scouts to Florida where sightings had previously occurred. During that year the Indian population were more helpful. More constructive attention was given to their new reports. Popular belief has always been that Indian information must be treated with caution, as their histories are steeped in legend. These are often quoted, and the legends range from tales of water-demons to semi-human giants of fabulous character. Like many legends among the unsophisticated, they are rarely substantiated, but they die hard.

The more recent Florida situation was different. The Indians involved in 1974 were reputed to have some integrity in outlook and experience. Their reports of tracks and sightings convinced the new research groups that South Florida had its resident Yeti/Snowman/Bigfoot, i.e., Relict Hominid specimens, even if attempts to unveil facts had to proceed warily. For however reliable these Florida Indians were reputed to be, even the most honest local informants are sometimes tempted to tell enquirers what they think they want to hear.

Nonetheless as examples that the unbelievable must sometimes be believed, let us remember fables which were eventually proved to be true, as in the case of the African Gorilla, until the 1880's thought to be a dubious traveller's tale; the Komodo Dragon of Asia discovered to exist in 1912; China in 1934 producing evidence of the Giant Panda; not to mention the later reappearance of the Coelacanth.

Local Florida descriptions by the Indians suggested that Bigfoot there equated in behaviour in many instances with the creature's alleged habits in other venues. A student of unusual animal and human distributions told me that the Red Man of the Americas had possibly been trying for a long time to tell those who would listen about something that might eventually be of great importance in the natural sciences.

Bigfoot everywhere is noted to communicate with whistles, screams, and hooting noises. Some searchers believed the species lived in small family units, but that territory lines seemed to overlap at times. Sometimes, Bigfoot played with stones or small rocks. He liked to crack stones together in a sort of set cadence. Here, there seems a link with Zana, the wild hair-covered animal-like female captured by hunters in desolate mountain forests of Zaandam, or near the seacoast of an area known then as Adzharia. The actual place and nature of her capture was vague, but she was sold from owner to owner, untamed, covered with heavy black hair, and eventually became the possession of a rich landowner, one Edgi Genaba[1]. She was described as an Abnauayu, which, means, "forest man", or "woman" in this case, if one can so term Zana. The language in that area was Abkhazian deriving from a Caucasian region. Zana's extraordinary story is told in my last book on this subject, "The Yeti". My reason for quoting her here is for one of those comparisons I have previously described. Because after she had been tamed a little Zana used to squat on the stones endlessly, much as the American Bigfoot is said to do. We shall meet Zana again later in this book. Something of her mysterious origin might be a little clearer, though only based on a tentative theory of my own.

In addition to the Florida survey, Peter Byrne's scouts from Oregon research station, covered areas of Oregon and Washington where reliable reports of sightings had occurred. They continued to recur at intervals. These investigations were in addition to the Florida ones. The country searched is a vast expanse of wilderness largely uninhabited by Man. It extends from Northern California north through Oregon, Washington, and Western Canada. The flora and fauna represent similarities with other terrains of alleged Yeti/Bigfoot distribution and extends over areas of the Cascade Mountains Range. It is a very rich land in species, and there is exuberant growth of vegetation.

John Napier[2], the primatologist and author, and others in this field, have asked: Is the Pacific North-Western forest's huge area in wintertime a life-maintaining habitat for large animals such as big plantigrade biped primates?

---

[1] https://dna-explained.com/2021/07/30/the-origins-of-zana-of-abkhazia/
[2] https://en.wikipedia.org/wiki/John_R._Napier

There is a plant and smaller animal reserve as possible food sources for Bigfoot. This northern-western wilderness consists of rocky meadow and hill grasslands, bordered on one side by a forested ravine. This is an extension of the great North-Western Forest. Grass and plentiful edible herbs and roots abound. These are highly lush, prized by the Indians and are relished by grazing and rooting animals. The supply is limitless. Most primates eat pasture plants, though they are carnivores as well as herbivores. Our central figure has an extensive and varied natural larder at hand under the ravine's thick canopy of oak, pine and Douglas Fir. Burrowing animals have food caches which are often raided by larger animals.

Even in winter the huge forest is the richest provider of food and refuge. Riverside logging does take place, but mostly on the terrain's fringes, so scattered timber work would be no problem for whatever lived within. Nature can thrive undisturbed except for rivalries and battles for survival between its own species of coyotes and lynx, deer and cattle, while the Golden Eagle reigns on high as a native predator.

Gorges and heavily wooded outcrops can provide lairs for Bigfoot, and so account for his unpredictable appearances and disappearances.

Detailed accounts of similar observations were recorded in 1884. The geographical background was thousands of miles away but offered the same distribution of plant and animal life. These were described by Joseph Dalton Hooker[3], M.D., R.N., F.R.S., in his "Himalayan Journals"[4]. He wrote how local tribesmen feared a mysterious race of wild creatures called the "Harrum-Mo". They were animal-like, covered with hair, and dwelt in desolate mountains and forests. They had near-human yet animal-like traits in appearance and habits. Their movable distribution was in the Tibetan borderlands. The region, like the Russian and American and Canadian wilds, was full of inaccessible mountains sliced with wooded precipices that formed a huge valley. The Harrum-Mo equated at some points with the Yeti and Bigfoot image, and comparatively civilised communities gave them a wide berth.

Sometime after Hooker's Journals, there appeared another explorer, William Woodville Rockhill.[5] He gave accounts of equal creatures in other parts of Tibet. Some Mongolian merchants, speaking of their forays for medicinal plants, told how among the wild cattle they encountered were creatures named locally as the Geresun Bambursh (The spelling translation of the last may be hazardous). These

[3] https://en.wikipedia.org/wiki/Joseph_Dalton_Hooker
[4] https://www.gutenberg.org/ebooks/6478
[5] https://en.wikipedia.org/wiki/William_Woodville_Rockhill

were wild men, the merchants said, hairy, erect, eating snakes and insects, but could not speak, and were more beast than human.

And those indirect references to the Anthropo-Zoology phenomenon came several years after the Himalayan tourist, Fraser, in his 1820 book, wrote of the "Bang's" footprints.

Study of the phenomenon in scattered references comes to us in the 1950's, 1960's and in the 1970's onwards. They are also very fully covered for certain areas in the works and investigations of not only Professor Boris Porshnev, but by Professor Bernard Heuvelmans[6], the late Ivan Sanderson[7], and the documentations of Doctor Osman Hill, formerly Prosector at the London Zoo. Charles Stonor wrote "The Sherpa and the Snowman"[8], and Ralph Izzard[9] produced "The Abominable Snowman Adventure"[10]. This was about an impressive exploration when he and fellow searchers found tracks in the Himalaya, but no Yeti, though they brought back valuable fauna information, and specimens of rare plants. John Napier wrote "Bigfoot" about the Snowman angle in the Americas, and René Dahinden went to Moscow to confer with Professor Porshnev and his colleagues, taking with him impressive documents, photographs, and other equipment. This was in 1972 when the Russian experts of the Darwin Museum pronounced as genuine the film of the Female Bigfoot taken by the late Roger Patterson at Bluff Creek, California in 1967.

John Green, of British Columbia, wrote several documentary records of his findings and first-hand evidence, and one very long book in 1978. But his trilogy of journals before this, in the form of personal verbal and dated evidence from local witnesses is down to earth, factual, and never once exaggerated.

On the first occasion when I was in British Columbia, I asked him why consistent attempts to take photographs of sightings were never successful. He explained that researchers in obvious terrain had set up night watches armed with cameras and relevant equipment. No Sasquatch had ever come within range. This was not surprising in view of the Sasquatch's erratic and infrequent appearances in mountainous country or seashores seldom visited by anyone except for special reasons like obtaining visual evidence of the creature's existence.

---

[6] https://en.wikipedia.org/wiki/Bernard_Heuvelmans
[7] https://en.wikipedia.org/wiki/Ivan_T._Sanderson
[8] https://www.abebooks.co.uk/book-search/title/the-sherpa-and-the-snowman/author/charles-stonor/
[9] https://en.wikipedia.org/wiki/Ralph_Izzard
[10] https://www.goodreads.com/en/book/show/10561691

Also, whenever isolated fishermen or a hunter has seen the phenomenon, he is so taken aback, minus a camera, and if a hunter only has a gun which he dare not use because "it looked like it could be a real god-durned guy!"

The late Ivan Sanderson[11], author of "The Abominable Snowman in Five Continents"[12] formed "The Society for the Investigation of the Unexplained". It was formed several years ago in the United States, and Sanderson's widow is still very active in the Society's work. Allen V. Noe is the Principal. Sanderson had been a lifelong researcher in the natural sciences. He was also a persistent and diligent Snow man investigator all his life. His Society to the day remains a museum of unusual occurrences and some of the phenomena's visible proofs, according to the protagonists of an out-of-the-ordinary cult. Allen V. Noe is convinced of the reality of the Snowman. Like many other searchers in this question, Noe believes there are even more sightings than reported, but that witnesses do not always speak of what they have seen for fear of being laughed at. This is always at the back of the public's general reluctance to speak. From the Society's documents come the repeated comment from witnesses who describe how the creature's eyes glow in the dark. Some say this is an orangy-red, others quote a golden colour. An animal characteristic which puzzles serious investigators.

One of Ivan Sanderson's correspondents over many years has been Gordon Creighton, who was of the Permanent Committee on Geographical Names. They exchanged letters often on topics of mutual interest. In one of these Sanderson conveyed an unusual point of view on the nature of the Yeti, though his idea was merely exploratory, and not conclusive. He was saying that it was strange that no camera had ever succeeded in capturing even a glimpse of the eluding creature. The two men's exchange of ideas went back and forth at intervals across the Atlantic, and in one letter Sanderson cogitated whether the Yeti problem might have supernatural overtones. That the inability ever to have caught even a poor impression of the sighting might be caused by camera-blocking at the very moment of sighting - some unknown influence causing a kind of extra-terrestrial discouragement! The truth about the Yeti never to be divulged.

The Yeti a phenomenon from another dimension? Could this be a message that Man was not intended to solve the riddle? Sporadic sightings could possibly merely represent the occasional human awareness of other-worldly entities? Sanderson wondered whether these appearances could have emanated from an atmosphere

---

[11] https://en.wikipedia.org/wiki/Ivan_T._Sanderson
[12] https://www.goodreads.com/en/book/show/223404

less dense than earth's atmosphere and were made known to humans only in rare moments.

This viewpoint prevails in some quarters, and this refers to circa 1983, etc.

Nobody knows of Ivan Sanderson's last postulation on the Yeti problem, but he might have spoken of this idea to others as well as to Gordon Creighton. Sanderson seems to have commented on this to Creighton not as a hard-and-fast opinion, but as a thinking aloud he had transmitted to paper for his correspondent's consideration.

My own mind is reluctant to accept an other-dimensional theory. Such a theory could not explain the footprints, even if there has been a record of unspecified footprints discovered and unidentified for centuries. And materially, there is the Female Bigfoot Film which certainly refutes any non-physical theory, and which is now considered evidence of a phenomenon to be assessed more seriously for the first times since 1967. But there again there are now enlargements and clarifications of frames; details developed from the Female Bigfoot astonishing movie which pose questions.

Civilisation is afraid of the remaining untamed wildernesses. It hopes that all earth and its creatures have been charted and catalogued. While the human race's leaders in influential quarters of so-called progress are bent on exploring the outer spaces of the universe, they prefer to ignore that some forgotten species might still be living sparsely in the enormous unvisited spaces of our own planet. In some of the mountains of Asia, in the far Russian boundaries, and beyond other borders of civilisation, north, south, east and west.

Real truth can only be found if all persons pool their information and experiences so as to re-explore without rivalry, and without too much modern equipment, in the relevant places. And this could include the non-physical.

Three speculative points occur to some researchers, if not to all of them:
1. Is the Yeti a psychosomatic phenomenon?
2. Could the Yeti be an expensive gigantic hoax?
3. Are we flogging a dead or dream Yeti?

I say No to all three questions, and will qualify them by quoting Ivan Sanderson's widow, Mrs. Sanderson, who continuing her work for the Society he founded, is the editor of "Pursuit" the Society's Journal.

She stated that there were three major explanations put forward to account for Yeti, or Bigfoot, reports. The first holds that reports are the result of misidentification of known animals or are simply frauds or hoaxes. The second explanation in her own words, seem "...to be the consensus of most students of the subject, that the creatures are indeed real, and in most cases are well known to the indigenous 'natives', but that they have merely avoided being caught up to date. The third view is that the reported creatures are the product of some paranormal, or extra-terrestrial agency."

Mrs. Sanderson noted that it was interesting to realise how adherents of all three theories recognise that no specimen has been captured, despite considerable effort. Each group explains this failure so far according to its own ideas on the nature of the beast.

(One could challenge her here over what seems like a certainty in applying the term Beast to the phenomenon).

The unbelieving first group say that one cannot capture what is non-existent. The group that asks those who are interested to wait for a while seems the most sensible. As for group three, she commented that they appeared to believe that the phenomenon by nature is so far removed from our usual understanding of the world "Animal" and "Capture" that securing one is a practical impossibility. This last viewpoint is too involved and vague to have any value.

Mrs. Sanderson thought that most Yeti hunters were firmly of the opinion of the group advising to wait and see. They explain also why the capture of such a large extremely strong; extremely swift, and probably intelligent creature, is a most difficult task. Any idea of hunting one with an extensive caravan of land rovers charging into the brush is sheer "adventure film" imagination and would be the wrong approach. The genuine exercise must encompass knowledge of the creature's habits and haunts, and involve great patience and perseverance, quite apart from possessing the necessary human equipment for temporary trapping.

Madeline Slade, known as Mirabehn, said much the same to me several years ago when we were discussing the enigma of the Himalayan Abominable Snowman, or Yeti. She called the mystery "Our lost brother of the mountains".

Chapter Nine
The Legends, The Bizarre, and The Related.

Bhutan is a northernmost Himalayan kingdom. It is of breath-taking beauty with its spectacular scenery, variety of flora and fauna, and its legends of enchantment and mystery, which include the Yeti.

Doctor Rammamurti was ceded to Bhutan a few years ago by the Indian government to help the King of Bhutan in the reorganisation of the country's unpredictable postal services. The wild mountain country with its river torrents, jungle-covered heights, and precipitous gorges had made communications difficult to maintain without recruiting modern advice. This the Indian official provided admirably and performed a hard project of modernisation satisfactorily. The good doctor had imagination as well as business-like drive, for he was interested in the country's traditions too, and took infinite care to give me news, pictures, and the legends of this aloof mountain land. This included the people's acceptance of the Yeti as once a fact. Their literature of as recent a time as 1966 stated that Bhutan still believed in the Snowman's existence. Many yak herders claimed to have seen the creature, or to have found its tracks when crossing high snow-bound passes.

Few Snowmen had been seen by the present generation. Many thought that though legend it might be, and possibly as extinct as a species, there was yet truth in belief in it as a real entity. So, in 1966, Bhutan commemorated the Snowman in a series of rather fascinating coloured stamps. They established it or him, as their national animal, legend though it may have become.

The last person of note to have seen the Yeti was reported to have been a sister of the then reigning King of Bhutan when she was travelling at a high altitude across the mountain passes a good many years ago.

The Snowman appears in some pictures on old Bhutanese religious scrolls, as well as on Tibetan ones. There were long descriptions of the elusive creature, its habits and appearance, and accounts used to be given of its occasional attacks on humans and yaks. Much of this is hearsay, and relevant direct sources are difficult to find. Documents list the three distinct types of Yeti: The large fairly docile one; another, a savage carnivorous specimen, about five fee tall with long hair all over, and apelike and muscular in build. Last comes the "little man" Yeti, shy and shaggy. Characteristic to all Yeti reports wherever their source, is the strong, pungent smell they exude, and their high whistling call. As stated before, that high whistle is not their only vocal sound, as at times screams and powerful roars have been reported.

Another typical story of the country I found was about a quarrel purporting to have occurred between two old animal friends, the yak of the high pastures, and the buffalo of the plains. Though the association with our subject is the rather tenuous one of the yak very occasionally being a yeti's victim, it is an agreeable mountain story, so, no apologies.

The tale describes how the Yak is always looking down on the plains, and the buffalo always looking up at the mountains, as they are both sorrowful over the quarrel which had separated them. They had probably forgotten what trivial thing had caused the argument, quite a human analogy, and must have occurred to one person who related the remorse of the two beasts in rhyme:

The Yak and the Buffalo
The Buffalo yearned from the plain
To the mountains where lived the Yak,
And the Yak looked down with regret
And wished that time would swing back,
For once they were friends, "For ever,"
They'd say in duet together,
"We'll never fall out, no never,
In snowfall, storm, or fine weather."

But a jealous demon made them
Have an argument,
And it blew up and down to quarrels –
Their parting was permanent
For the Spirit of Quiet of the mountains,
And the just God of the plain,
Ordered a long repentance
Before they could meet again.

"Oh Yak, Old Yak, I miss you
of the shaggy lumbering walk,"
Sighed the Buffalo in sorrow
At loss of their happy talk.

"Buffy, you dear old soppy,

Of slow and rolling tread,
And your gentle, cheerful cropping
Where together we'd be led.
I look down at your grazing,
My own appetite I've lost,
The chaps around don't understand
What our separation's cost

And that is why the Buffalo
Always with mournful eye
Looks up to the snow-tipped mountains
Where his friend lives near the sky.
And the Yak with lowered head and gaze
Stares sadly down below
To where his friend is hoping
To hear a celestial, "Go
Forth, make up the quarrel
On the pass where spirits ride,
Where grazing is green and forever
And the gods of peace abide."

Genuine references descending from ancient or mediaeval ages at times can become confused with the recorded fantasies and self-styled experiences of certain long-past adventurers. Though these stories should be kept apart from histories that have substance, some for good reason have found a place among the classics of literature.

Journals of fabulous travels used to be popular forms of wonder and entertainment. The only wonder they might occasion now is at the audacity of their perpetrators' fables, as such chronicles often were, but some of the entertainment value remains.

A case in point is that of Jean d'Outremeuse[1], born 1330, a scribe and chronicler of Liege. His serious labours must have been boring. Perhaps it was to vary the monotony that he took advantage of his undoubted literary skill and his historical studies to produce "Sir John Mandeville's Travels"[2]. This knight may have existed, possibly under another name. d'Outremeuse no doubt culled references about a somewhat similar flamboyant character during his research work, and produced a wandering fantastic adventurer who as a soldier of fortune scoured the globe for experiences and discoveries in strange lands among stranger men and animals.

[1] https://en.wikipedia.org/wiki/Jean_d%27Outremeuse
[2] https://en.wikipedia.org/wiki/Mandeville%27s_Travels

D'Outremeuse's feats of imagination can be ranked with those of Munchausen. Reputable and respected men of letters and scholarship later on researched diligently on the scribe of Liege and his literary, personal, and historical resources. They decided that he was a clever and inventive rogue.

It is possible that d'Outremeuse was not a deliberate trickster but had concocted the travels of the debateable Sir John Mandeville as a private joke to relieve the tedium of his days. He may have enjoyed watching the sensation Sir John's adventures caused among the more gullible among Flemish, French an English readers.

Some of his sources of information did derive from accepted records, as for instance in a chapter in the "Sir John" saga about one Friar Odorious. This character, a monk of some learning and reputation, did exist once. This particular chapter is called "The Journal of Friar Odorious". It deals with the Far East and India and describes how in his wandering ministry the holy man was once invited to visit a temple in India where a strange scene was enacted for his benefit.

In the language of those times, "Sir John" describes the friar's temple experience, quoting the monk's own words: "... a Monastery where many strange beasts of divers kindes doo live upon an hill."

One of the religious men of the abode took "Two great baskets full of broken reliques which remained on the table, and led me unto a little walled park, the doors thereof he unlocked with his key, and there appeared unto us a pleasant faire greene plot into which we entered. In the greene stands a little mount in form of a steeple, replenished with gragrant herbes and fine shady trees. And while we stood there, he took a cymbal or bell, and range therewith as they used to ring for diner or bevoir in cloiseters, at the sound thereof many creatures of various kindes came down from the mount, some like apes, some like cats, some like monkeys, and some having faces like men. And while I stood beholding of them, they gathered themselves together about him, to the number of 4200[3] of those creatures, putting themselves in good order, before whom he set a platter, and gave the said fragments to eat. And when they had eaten he rang upon his cymbal a second time, and they all returned to their former places. Then wondering greatly upon the matter, I demanded what kind of creatures these might be. They are, quoth he the 'Soules of noble men which we doo here feed, for the bve of God which governs the world, and as man was honourable or noble in this life, so his soule after death entereth the body of some excellent beast or other. But the soules of simple or

---

[3] This figure is debateable

rusticall people do possess the bodies of more vile and brutish creatures. Then I began to refute that foule error, howeverbeit my speech did nothing at all to prevail with him, for he could not be persuaded that any soule might remain without a body".

So, Friar Odorious, finding the good monk too fixed in his peculiar convictions on the afterlife, went on his way, no doubt to convert more malleable human material.

The account of this so-called ancient semi-zoological gathering, "Some with faces like men", strikes a note of Yeti reminder, and Friar Odorious conveys the impression of being a piously consistent character even if he does figure in "Sir John's" adventures, and his ponderings seem coloured with Romance challenging Reality.

Many might experience strange encounters if the mind could regress back far enough and consciously among the branches of the tree of life, as expressed in some of the phases disclosed in "The Stanzas of D'Zan", by Madame Blavatsky[4].

Echoes of the Snowman, or monster image sounds through its pages. These stanzas are a collection of allegedly very ancient words of wisdom, warning, and ethics. They reach back to "The Secret Doctrine". These chronicles might have penetrated Western thought from the pages of the Doctrine. Madame Blavatsky was an adept protagonist. Tangible elements derive from the presentation of these strange sayings. Stanzas could have been inspired once by Sufi, the name by which Islamic mysticism became known round about the 9th century, A.D. The Arabic term "Sufi" meant a weaver of wool. Woollen dress was associated with spirituality even in pre-Islamic times. Sufi is being of oneness and is expressed also in some verses in the Koran.

However, whatever was the original inspiration of the "Stanzas" some peculiarly Biblical pronouncements emerge from them in which humans, animals, and sub-creatures equate in some respects with the Snowman riddle.

One passage is headed "The first sin of the mindless men". The text runs: "And those who had no spark, took high she-animals unto them. They begot upon them dumb races. Dumb they were themselves, but their tongues untied. The tongues of their progeny remained still.

[4] https://en.wikipedia.org/wiki/Book_of_Dzyan

"Monsters they bred, a race of crooked, red-hair-covered monsters going on all fours. A dumb race to keep the shame untold."

"(Race of monsters). Not the anthropoid or any other apes, but verily what the Anthropologists might call 'The Missing Link', the primitive lower man, the pseudo-man, not the real man. Even this race will find itself on one of the seven paths on the last day".

Those Seven Paths refer to progression in the religion of Buddhism. What pulls one up short in this enigmatic literature is the reference to the Missing Link. Certainly not a topic for speculation in the days of antiquity the Stanzas were supposed to represent. And the red-hair-covered monsters do suggest a Yeti connotation.

Chapter Ten
From Antiquity Downwards

The Yeti appears intermittently in pages of antiquity. The image appears under various terms and against various backgrounds. References to its nature as a near-human species is not confined to Bhutanese and Tibetan art and some traditional folklore.

Several years ago, Emanuel Vlcek, M.D[1], a professor of the Czechoslovakian Academy of Science in Prague, went on an expedition to Mongolia mounted jointly with colleagues. Among the agenda projects was the quest to discover anthropological and medical data from Mongolian and Tibetan literary spheres. The doctor studied Tibetan books from the library of a former Lamaistic university in Gandan. The university did not exist as such anymore, but he found a book written by two authors of long ago called Lovsan-Yondon and Tsend-Otcher entitled "Anatomical Dictionary for recognising Various Diseases". It was printed originally on long narrow strips, characters appearing on both sides.

Apart from medicines and prescriptions quoted, the authors had given a list of the Fauna of Tibet and neighbouring regions including Mongolia. Illustrations showed panels depicting natural history of ancient times. Through later research it was found that these were copies of mural pictures in temples of the past. They showed groups of monkeys, quadrupeds, birds, and insects. The yak was depicted too, but among this very old bestiary of identifiable animals was one picture showing the wild hairy man. A distinct contrast and a definite species all on his own.

This dictionary had had more than one printing, the latest being at the end of the eighteenth century. It was produced again many years later in Mongolia. The Hairy Man was shown as very hirsute except for his hands and feet. Names attached to him were: Sandja (Tibetan); Bitchi (Chinese); and Kumshin Gerusgu (Mongolian). In later editions of this work the Wild Man is called Osodrashin in Tibetan; Peevi in Chinese, and Zerlog Khoon in Mongolian. So, one realises here again that there used to be more than one specific word-sound for the phenomenon Man-Beast or Beast-Man. But all such names meant "Man Animal". A Tibetan text to one of the illustrations stated: "The wild man lives in the mountains. His origin is near to the bear. His body resembles a man's. He has enormous strength. His meat may be eaten to treat mental diseases. His gall cures jaundice", Other records[2] quoting long-

---

[1] http://alamas.ru/eng/publicat/Vlcek.htm
[2] Professor Boris Porshnev (Quoted in "The Yeti" by O.T.)

past days describe how in remote regions of Russia and Mongolia the wild men were hunted for the medical use of their flesh and organs.

The question could be how much fact can be gleaned from these scattered reports from antiquity. There must have been at least a scintilla of accuracy, or these pictorial and written records would not have been made, not would their nature have persisted elsewhere.

It may not be possible to deduce accurately how far back removed from the Neanderthals this man-animal would have been, or even if he ran concurrently with them, for the Neanderthals by comparison with this type seem humanly civilised.

The ancient hairy wild man may be a few cousinship's removed from the Yeti, but he is certainly one of his fellow-travellers.

This enquiry would be incomplete if Carl Linnaeus, the great Swedish botanist and naturalist was not included. Among his extensive works his 'Systemae naturae' contained an analysis of certain obscure species that qualify for reference here. In his immense research and studies Linnaeus had collected detailed accounts of certain primitive sub-human types. This work was maintained at times in collaboration with colleagues who travelled far afield in research for his investigation.

Carl Linnaeus was President of Anthropomorphia, and responsible for this publication's themes. One of his papers a Latin translation made a statement dated Upsala, September 6th 1760, which said: "This is submitted by Christian Emanuel Hopius.... After discussing various types of monkeys and apes (Simia) namely, the African Pygmy called Pygmaeus, the Satyrus, also from Africa, the Lucifer from Java and the Nicobar Islands, comes finally the species called Troglodyta, or Homo Nocturnus".

The writer was referring to Systemae naturae where Homo Nocturnus is described, and his report continued: "These are children of the night, for whom day is night and night is day, and in my opinion, they are very close to us. Ever since the days of Pliny the name has been noted, and in such regions as Ethiopia, Java, Amboina, Mount Ophir in Malacca, and elsewhere, often dwelling in subterranean caves."

In the original Latin version, there is a footnote in which Pliny announces: "Troglodytes excavate caves. These are their homes; their food is the flesh of snakes.

They cannot speak, but make a screeching noise, and have no proper language or conversation."

One Bontius[3] is quoted too in the documents and he describes Java experiences on the basis of Homo Nocturnus being a living specimen. He says of the species: "I saw several of both sexes walking about erect, particularly the female covering herself in much shame from the eyes of unknown men. And also covering her face with her hands and weeping copiously, uttering cries, and copying the actions of other humans, so that you would say that, apart from their lack of speech, there was nothing human lacking in them. The Javanese however say they can talk but do not desire to. Nor are they able to get them to work."

The documents amplified: Homo Nocturnus's dimensions were little more than those of a nine-year-old boy. Their colour was white, probably because they were never burnt by the sun since they only wandered about by night. "They go about erect just like us, and their hair is short and curly by nature like that of the Moors. Their eyes are round, and the pupils and iris are golden colour, a point that calls for particular note in them."

It is at this point, centuries later, that we too must take particular note of this peculiarity referred to earlier. The gold in eye colour is a reminder of the American and other hairy creatures described as having self-glowing eyes in the dark. The Linnaeus paper added: "The eyelids are drooping in front, whence the fact that they have to see obliquely or sideways. Beneath the eyelids they have a blinking membrane, after the fashion of bears, owls, and other creatures want to be abroad at night, which feature at once sets them apart from us."

In the continuation of the Linnaeus document another voyager of those days is quoted. He was from Belgium and had returned there from the East Indies and said he too had seen strange beings in Java. His account agreed with those from other travellers. Another tourist of those days spoke of his visits to Northern India. He had just come back. His travels were in the regions of India from which had started the first Abominable Snowman references in the English language during the 19th and 20th centuries.

He spoke of similar species, having been told about them while not actually encountering any as the Linnaeus researchers had done. There was a description of a creature with arms much longer than human arms, and "The fingers of their hands

---

[3] https://en.wikipedia.org/wiki/Jacobus_Bontius

reach to their knees...." This is in keeping with the reports of two continents. Their narrative went on: "They conceal themselves in their caves by day in order not to be seized by men, and are almost blind, and feel their way with their feet. They have to get accustomed to light. At night they see clearly. They attend to their affairs in the dark and hide from men whatever is going on and whatever they think relates to the advancement of their own species. For which reason the inhabitants of those regions kill and slaughter them most mercilessly, as if they were most evil thieves whenever they come across them." The narrative there had presumably reverted to the Java distribution of the strange species. "They have their own language," said the report, "which they speak with a whistling noise, and which is so difficult that it cannot be learnt except by long habitation among them. And for learning our language they are quite inept, being unable to say anything beyond the affirmative and negative...

"Certain writers say that these beings claim they were driven out by men, and that now they harbour the hope that the time is at hand when they will regain the dominion that they once had, though this day never comes..."

One part of the Linnaeus data gives a footnote by one Dalin - probably another of those early wanderers in search of strange lands and stranger people who states: "In Central Africa a species of snowy white man is found, whose hairs are white and matted, their ears long, their eyelids dropping, eyes round with a red iris, and having a transparent yellow membrane on the pupil. They look sideways in either direction at the same time and see better in the dark. They live to the age of 25 years. Their bodies are small and thin. They also say that the earth was created for their benefit, and they hope to secure again eventually the dominion over it."

According to other pages in this documentation, people in many places in the East Indies used to employ sub-humans for performing easy household tasks such as carrying water or bringing and removing dishes.

There is an account of how one Homo Nocturnus was captured out at sea and brought aboard a ship where he refused to eat cooked food or look at the light. When he walked, he would lift his feet high up on the deck's unfamiliar level floor as he used to do in his native woods. The short anecdote does not explain how this poor Homo happened to have strayed so far from his comfortably obscure forest.

The Dalin Central African reference to the snowy white wild man calls to mind the legend from "The Epic of Gilgamesh" where Enkidu, the savage animal-loving wild man, became as a brother to King Gilgamesh, as they venture in wildernesses together searching for truth and their fates. Enkidu forsaking the drinking pool where he and the beasts of the hills and deserts foregathered before he became a real man and King Gilgamesh's loved brother.

From the summing-up of the Linnaeus's accounts, it is fairly clear that his Homo Nocturnus equates in varying degrees with the Relict Hominid of many names of Russia, Mongolia, and Tibet, and also with the Snowman or Yeti of the Himalaya. Also, the Bigfoot/Sasquatch puzzle of the Americas.

Some of the more remote Moslem communities of Russia and Asia used to call the Yeti or Snowman the Shaitan, meaning a demon entity. Derivations of the term, Shaitan, that is, Satan, appear in various parts of the world. One often finds widely world-separated terms that sound the same, but that academically they are quite unrelated.

I found a world-cousin to Shaitan thousands of miles from its Eastern source. Some years ago, an observation survey was taking place in what was then Tanganyika, now Tanzania. This was a mapping process, and the man in charge had sent some of his team with heliographs and other equipment up a mountain called Genda-Genda.

After a few hours the African workers came down and refused to climb the peak again. Their employer asked them why, and they replied that there was a bad spirit up there, one Shaitani, or Shitani, the Swahili sound.

Fellow Travellers are word travellers too.

Chapter Eleven
In Hairy Company

Only by their one peculiarity do the characters in this chapter justify their inclusion. For among the Fellow-Travellers they are the most unique: The humans separated from normal humanity by their extraordinary physical defect.

This defect was studied for a lifetime by two French doctors of nearly one hundred years ago. Had their studies caused as much concern among traditionalists as once did the Darwin theory, their names might have echoed around the world.

They were Professor A.F. Le Double of Tours, and Dr. Francois Roussay of Pont-Levey. They were not concerned with any ape-men ancestors, but with actual living humans bodily covered with luxuriant hair. They are mentioned briefly in some reference books, and one of their publications, "Les Velus"[1] which means "The Hairy Ones" was tracked down at Toulouse University after a long search because it suggested some indirect Yeti relationship. "Les Velus" appeared in 1912.

Le Double and Houssay had applied immense research and analyses on the excessive hairiness they had found in certain human species. They were deeply absorbed in these primitive hirsute specimens, and one speculates on what happened to their practices during their fascinating lifelong studies. So, their research progressed from one to another of their discoveries in heredity's physical inheritances that led to deviations and the "sport" freak elements in descendants.

One young girl in Germany was sketched in a 1493 journal, "La Chronique de Nuremberg". She had a pretty and normal face, and shoulder-length fair curls, but her whole torso and limbs grew long, thick blonde hair. Another old document the industrious doctors had unearthed, showed a hairy man with a lion-like face that was hardly visible for hair, which also grew all over his body too. This paper was named "La Traite de Rumania Physionomie de Porta".

By one of those pointless but odd coincidences that occur in records from time to time, "Les Velus" was published during the same year that Science on this side of the English Channel acclaimed the Piltdown Skull, later to be denounced as a hoax. There was no invention about Le Double and Houssay's hirsute men, women, and children. They located them far and wide, and collected their life histories and developments in the interest of medical science.

---

[1] Les Velus", 1912, Vigot Freres, 23 Place de l'Ecole de Medicine, Paris, now only available at Toulouse University.

A sample of hirsuteness from the Far East occurred in Krao, a six-year-old girl. She was brought to Britain in 1880 by a well-meaning traveller. The child was covered with a growth of strong black hair. Her face was more like a gorilla's than a human being's. When she was twenty Krao's patron must have lost interest in her. The showman, Barnum, took her up, and exhibited her all over France, where one loses track of her.

Schwe-Maon of Burma and his daughter, Maphoon, figures next in the hairy creatures discovered. One John Crawford, friend of government local officials, befriended the Schwe-Maon family. The freak deviation derived from the masculine line. They did not show ape-like features but were covered from head to feet with thick silvery hair which would have been attractive had it grown only in the right places.

Maphoon actually married. A dowry had been given to her by her benefactors, so this was a help. She produced two sons. At first the little boys seemed normal jolly little Burmese, but they became heavily hairy as they grew up.

Le Double and Houssay diagnosed the abnormal hirsute condition as Hypertrichonique, a term which speaks for itself.

The "Les Velus" journals switched from the far East to France's South-West. In the ancient times of this mountainous region, it occurs to the researchers that there could have been a pact of "live and let live" between stone-age man of the Pyrenees and local animals. These humans of antiquity often had to share cave shelters with bears and other carnivores, however risky. But probably the human/animal link was closer then, than now. As glaciers retreated better caves offered safety of a kind and warmth, so that primitives could avoid the lower creatures whose tempers depended on what they found to eat. This was safer for the ancients, whether they were hairy sub-humans or not, though hairiness was probably desirable through lack of central heating.

To their studies of hairy types, le Double and Houssay brought a look at the descent of man. They found quite a lot of whiskers there. They gave an account of "The Man of the Caves" of Chokier, near Liege, Belgium. They discovered a painting of a stone-age hairy family against a primaeval forest background. Gabriel Max[2], the artist, depicted a male, female, and small child in arms. This was round about 1894. The picture was underlined with the caption, "Pithecanthropus erectus alalus - Primitive

---

[2] https://en.wikipedia.org/wiki/Gabriel_von_Max

Man half ape, half man". In their own documentary the two doctors added their own quote: "From which humanity derives". They also informed that one Jean Bourdichon[3], a scientist of four hundred years earlier, had expressed similar views.

Le Double and Houssay allowed themselves the indulgence of quoting one hair-slanted story which is pure legend: In the words of its old-time unknown writer: "... Mary Magdalen was of exceeding beauty, and when the Lord died bodily, she retired into a desert to expiate for her past sins even though Jesus had assured her she was forgiven. Soon the cruel burning sun destroyed her clothes which fell away, and she was naked. So, God in His mercy gave her a beautiful abundant head of hair that veiled her limbs from, the intemperate sun beating down on her self-chosen wilderness".

Note: Pietee, an erudite explorer in archaeological an anthropological history, discovered many years ago, the cave known as Caverne du Mas d'Azyl. He found artefacts and rock drawings. These included half human stone-age art. One scene showed an early, hardly human male among the animals. This nearly human form representing a living entity, suggested it was more Man than Animal: The Anthropomorphous neighbour of Pithecanthropus, and nearer true Man than the apes up till then had been known to be.

---

[3] https://en.wikipedia.org/wiki/Jean_Bourdichon

Chapter Twelve
African Fringe

The African mystery creature or legend occurs more in the Eastern areas of the continent than in other regions. It has more puzzling and contradictory facets that appertain to its partly equivalent mystery of the Himalayan, Russian, Canadian, and American backgrounds. These variances make it more perplexing and, in some cases, somewhat removed from the classic original image.

From the beginnings of investigations in the beast, these were always finally rejected as a doubtful mystery by zoology and science after examination. Such a result was not surprising. Not only in view of the wide, desolate and varied nature of terrains and the tribal evidence, but because of the superstitions which were rife in remote regions, and even where modern common sense had penetrated, if one can use such a challenging term in circa 1983. One can expect any kind of evidence from quarters where a mysterious and mischievous spirit known as the Tokoloshe would once be held responsible for broken marriages or an employer's broken china.

Talk occurred (1979) of "white" wild men living in hidden communities deep in East African wildernesses. They too were attributed with certain Yeti-slanted characteristics by some tribes a party of exploring scientists met.

Reports stated that the forests in part of the Mau country provided the source for many of the stories of encounters with the unknown creature. Sometimes settlers or game rangers or game wardens heard about weird beasts or creatures. The mystery being had different names according to area. Skins were seldom available, but on rare occasions when a hunter produced a pelt which he said was that of a Chemosit or Kerit (the two chief names) the officials who examined them generally found that they had belonged to an identifiable species.

A story of many years ago was told in Kenya of how the Wanderobo tribe shot such a so-called mystery beast with a poisoned arrow. It vanished wounded, not before it turned around and charged them aggressively. It was finally tracked and killed. Like many stories from forest or bush, no record exists telling how the tribesmen fought off the attack. Here, like in the case of critics of the Yeti reports, practical opinion challenged such stories with comment that no carcass was ever available for scientific examination except in isolated cases when then a skin was proved to be of a well-known beast.

Actually, when such incidents occur in wild country the lack of preservation is normal, evidence being soon destroyed by predators or climate.

If on occasions such skins were kept, it was by the vigilance of Army officers, Games Wardens, or other officials who sent the pelts back to Britain with the African label of Chemosit or Nandi Bear, to find that it had belonged to a giant spotted hyena. Liable to be a very big attacker, as this was realised once when such a skin was measured against me. It surpassed the top of my head by several inches more than five feet.

The so-called mythical name of the Nandi Bear had once gained a jokey reputation in some of the bars of Nairobi. A Nandi modern tribesman I met told me of its sinister reputation whether called Chemosit which means something wild, huge and unknown from the forest, or Keri which was taken seriously by the old people of his tribe many years ago. He demonstrated how they wore baskets on their heads when emerging from their huts, so that the Chemosit would not attack their heads to eat their brains. This is also attributed to some far away Yeti tales. The Nandi children at one time had a game rather like our own "Here we come gathering Nuts in May" in which the children are eaten up one by one by the hungry Chemosit, and then miraculously restored to their parents by a long-dead grandmother.

The giant spotted hyena is probably the most likely candidate for stories of the attacks by Chemosit or Kerit, though the baboon follows as a possible solution, especially to accompany reports of a mystery animal or creatures that at times walks bipedally. This is demonstrated in a true story of many years ago when an engineer on a rail construction site was attacked in his hut by such a creature and fought it off.

A feature that is similar in every land where the mystery occurs is that after an encounter, and the footprints are discovered and not identified with any known animal, if dogs are set to track the marauder down, they refuse to do so. No record, as far as is known, as to whether this happens in the Himalaya. Probably not, as such incidents are taken as a matter of course, accepted as accurate, or retold with occasional embellishments for the curiosity-element in uninformed visitors.

To return to the African scene, in my opinion the brains stories are superstition and legend, whether relating to the Yeti or the Nandi Bear/Chemosit/Kerit.

[1] https://www.abebooks.co.uk/book-search/title/the-heart-of-the-hunter/author/laurens-van-der-post/

Yet such myths are often supported by tribal beliefs. In Colonel Van der Post's book "The Heart of the Hunter"[1] a small link with the mystery creature-animal theme is detected in one instance. The bushmen, whose lives and lore figure largely in some of Van der Post's African writings, describe fate as their "Day of the hyena", a term evidently used to indicate the nadir of calamity.

Pliny the Elder's Encyclopaedia of natural History contained many of his findings of the year A.D. 77. One of his references speaks of the hyena in these words: "It simulates human speech calls out to shepherds in their homesteads and then tears them to pieces". (Vol.III. Books VIII-XI).

In Professor Bernard Heuvelmans' book "On the track of unknown animals"[2] it is postulated that the Nandi Bear was once an actual bear. Such a species has never been found in Africa within recorded memory. But in the long past days of Roman occupation bear fights were enacted in Numidia (now Algeria). In these cases, the bears for this Roman holiday arena were probably imported ones, but later circumstances could have been favourable to the animals leaving their traces. One cannot really know. The Professor quoted the name Duba as being occasionally applied to the creature. The term means Bear in some Arabic communities.

Many years ago, a young worker of the Nandi tribe was shown pictures of Indian jungle animals. Asked if any picture reminded him of the Chemosit, the boy without hesitation picked out the illustration captioned with the name Balu which is the Indian term for Bear.

Mrs. Mallett was a dedicated animal researcher in Kenya a good fifty years ago or more. She was a sort of very useful helping guest to a couple who ran a farmstead in remote country where there was good riding, a favourite pursuit of theirs. Her room opened out on to the compound from where she could view wildlife as she lay in her bed in the dark, and she had collected and cared for several small species that interested her. She was in the habit of taking walks into the nearby virgin forest, taking a ball of string attached to a firm tree trunk at the edge of the woods so that she would find her way back. She would play it out as she ventured into the labyrinth of trees and brush, and usually took with her a small African boy, the child of one of the farm workers. The M'toto, as he was known, would carry her basket and any other impediments.

[2] "Sur la Piste des betes Ignorees" Heuvelmans, Plon 1955. In English Hart-Davis 1958.
* RF - lemurs are found only on Madagascar not mainland Africa.

On one day she must have ventured deeper into the trees than she intended and was tracking down a small beast of the squirrel or lemur[*] species she had already spied earlier in the day. Gradually she became aware of a strong and increasing obnoxious smell, and the rustling and creaking in the undergrowth before her. There emerged a huge, shaggy, grey-furred mysterious animal, lumbering along towards her slowly and menacingly. It was definitely quadruped and like no animal she had ever seen in East Africa, the head round with ears small and close to the scalp. She looked around, but the M'toto had fled, and so did she winding up her ball of string in frantic haste, and stumbling and sliding over thorn and fallen branches until she reached the farmstead.

The M'toto was waiting in the compound expecting a scolding. "What was it?" she asked. He replied: "A Chemosit," and he ran away to his family quarters.

It was only in the 1950's that at last Mrs. Mallett allowed her forest experience to be published.

The consensus of opinion then was that the creature had been a giant forest hog in spite of its unusual appearance.

From other directions from time to time have come images of the dreaded predator as lionlike or of a fabulously large feline quadruped, or again as the upright-standing bear or ape type.

It must be remembered that the mystery image now takes a very back number in the areas where it was once referred to more often. The semi-legendary picture is not popular local history in the progress and modern development of fast emerging communities. A traveller could meet many people of a Nandi community, or of neighbouring tribes, and never hear reference to the semi-legendary creature. In such situations everywhere, much of folklore and local custom runs the risk of relegation as best forgotten and to-be-obliterated superstition.

But folklore and mysteries remain nonetheless even when they become unfashionable until they are seen and collected once more as interesting records of a country's evolution.

In the 1970's I received information in chatty letters from an agricultural scientist friend posted in Kenya after a study session in Britain. He sent me two reports from the "East African Standard" of those days which dealt with the identity of the creature he called "our old friend, the Chemosit". He said the letter reports he sent me might have some useful odd scrap. He was well acquainted with the relevant areas and their fauna.

He thought the letter purporting an encounter with a Nandi Bear probably referred to the spotted hyena and mentioned it as standing 4 feet high at the head. This would tally with the measurement once taken against me quoted previously, as the latter was taken from the shoulders to the hind quarters, probably making the full length, or height, more than five feet, since head measurements were left out. My correspondent ended his last letter with the exhortation of "good hunting" to which I of course reciprocated. So, wherever you may be, John MacFarlane, good hunting again. And here are those long-kept letter clippings: Just for curiosity's sake, and for the record:

The Nandi Bear
From "East African Standard" August 8th, 1966.

"The article by Mr. Lavers revived memories of my discussions on the subject of the Nandi Bear when I first came out to Chepair in early 1921. The consensus of opinion then was that under the pseudonym of the Nandi Bear the Africans called on the Kiri and the other the Chemosit.

From descriptions the Kiri seemed to be an enormous leonine hyena, whereas the Chemosit, which seems to have been the rarer of the two, was more bearlike and was greatly feared by the Africans. In those days several Europeans could report having heard weird cries which Africans told them were made by one or other of these two legendary beasts, and some reported having seen bearlike footprints.

Miss Clara Buxton reported that during a visit to Nyasaland, she had seen on the floor of a residence a skin that corresponded to the description of the Kiri. She was most excited about it and the owner agreed that it was the skin of a Nandi Bear. On urging him to send it to England for identification, he replied that there was no reason to do so it was just the skin of a Nandi Bear which was well known in Kenya. Nothing that she could say would persuade him that this was not so.

The Chemosit was reported to be a fearsome animal of immense strength who would on occasion climb on to a thatched roof, tear his way into the hut, smash the skulls of the inmates and eat their brains. He was addicted to rising and shuffling along on his hind legs in a bearlike manner, and it was from this characteristic that he was given the European name of the Nandi Bear.

A year or two ago in the Cherangani Hills, I was talking to an Elgayo and I gathered that he thought the Chemosit existed in the depths of the Cherangani Forest, though his name for it was different. I think that the legend of the Chemosit will always remain a legend.

Signed: F.H. Le Breton. Endebess, July 21st 1966.

There are two contradictions in that letter: The so-called Kiri of which Miss Buxton described the skin, the owner said it was the skin of a Nandi Bear whereas earlier in Mr. Le Breton's letter, the Kiri is presented as probably a huge hyena. The writer does not mention who Miss Buxton was or is.

From the "East African Standard" of August 10th, 1966, is a letter from a Nandi tribesman, Mr. Stephen Kipto Boen. He wrote:

"I have just come from Nandi where I was born, to work in Nairobi, and I was interested by what was written about the Nandi Bear. This animal has been talked about by old Nandi people. It was common in the old days.

As I was an eyewitness when one of these beasts broke into my neighbour's sheep hut, I came to realise that there was this kind of animal. I saw that it was of the Hyena family but not a hyena. It stood about four feet high, with long hair and with teeth bigger than a hyena's that could bite through a hut wall post.

When my neighbour heard the bleating of his sheep, he ran into the hut and saw that the beast was inside. He screamed and all my neighbouring people ran to see what happened. We found that the beast had bitten four sheep, mostly about the head. We speared it, but before it was dead it stood on its hind legs with its front legs loose, trying to hide from the people.

I noticed the big upright ears with a small head, as small as a leopard's. This made me believe what the old people said. They said that it waited for a person going along through thick forest or going out of a hut at night. It used to jump at the head, get hold of it, and run away, leaving the body behind.

That was when it was hungry but mostly it used to eat sheep.

The people were clever to distrust this beast. When they went out at night, they used to put small cooking pots on their heads. As the bear used to attack the head, it did not fail to take hold of the pot and run away with it, thinking it had taken the human's head. It was late in 1957 when I saw this beast. It exists, however, in the thick bush of our escarpment, and still our tribesmen near this escarpment complain that it attacks their sheep. It has two names. One is Kerit, but it is mostly called bour our Kenjin Mokololut. It is well known by these tribes up to today. Although it cannot be seen at daytime, this also applies to the Hyena, because these beasts live in caves in the thick bush and come out at night."

Some years ago, the B.B.C. commissioned me to research and take part in a programme on the Nandi Bear. Though already well away in Yeti study, the Nandi Bear was only a vague name I had heard of jokily. When I read my producer's terms of reference the name Chemosit hit my memory. I remembered how many years ago when I was very young, and I read a newspaper report about a French visitor to the Himalaya who was told a tale of a monstrous, frightening animal they called the Chemisette. And I laughed and laughed because in French this means a lacy piece of neckwear.

Only during the B.B.C. programme did the two names link up.

Of course, that old Himalayan garbled term derived from the East African term Chemosit. I still refer to it with a mental giggle.

When had it travelled such distances from one continent to the other?

On this note on an unsolved, unimportant linguistic mystery, closes this chapter about the Yeti's African Fellow Travellers.

Chapter Thirteen
The Chuchuna of Arctic Siberia

All reports of the Chuchuna of Arctic Siberia, a mysterious allegedly extinct species, come solely from a wild mountain area of the Verkhoyansk Range. It lies between the rivers Lena on the West and the Yana on the East.

This is Arctic Siberia.

Legends or accounts about the mystery tribe assert that they came originally from the East. Some say in the direction of Chukotka right on the Arctic Circle. A famous Russian explorer of the Far North, Vladimir Bogoraz[1], wrote an important monograph in 1914. In it he stated that the Chukchis of the Chukotka community, would tell of existence of tribes of giants ... A Chukchi drawing showed a giant said to be clothed in walrus meat. He came out of the sea, they said. He was very heavy and left tracks everywhere. One day he lay down to sleep, and three men saw him and caught him. They bound him with stakes driven into the ground. Then they killed him with their spear.

This is according to Vladimir Pushkarev[2], a geologist, who investigated the Chuchuna reports thoroughly. He travelled in the region with a companion, V. Pupko, a teacher in the Moscow Institute for Engineering Construction.

The "Giants" story smacks of legend, but many puzzling features relating to the Chuchuna of Arctic Siberia, are, or were, more fact and experience than mere fiction engendered by the desolation and harshness of the sparsely populated terrain.

Vladimir Pushkarev stated in 1975[3] how he and his companion visited the Verkhoyansk Mountains on the lower waters of the Lena and the Yana. There, some of the Chuchuna tribe were encountered as comparatively recently as 30 or 40 years ago, if one travelled with the reindeer herdsmen in the mountain pasture lands. In Pushkarev's translated report, the last syllable in Chuchunaa is written with a double a.

They researched villages situated on those two rivers, and joined the reindeer herdsmen of modern times who were still travelling in the ancient tradition with the

---

[1] https://www.britannica.com/biography/Vladimir-Germanovich-Bogoraz
[2] https://vk.com/wall-7240829_1698?lang=en
[3] His long paper "On track of a Legend" in VOKRUG SVETA ("Around the world") No. 6. June 1975.

herds to and fro on the summer grazing grounds high up in the mountains. They talked to men who had been keen hunters previously, and also discussed the Chuchunaa question with dedicated local persons steeped in tales of the vast area's memories and lore. These local people of the mountain villages by the rivers near Yakutsk, gave the two investigators realistic pictures of the Chuchunaa distribution. Gradually, their memories left the area of fairy tales to frighten children from wandering away from the home steads, and grew into a history that was real, even if now almost entirely of the past.

The last Chuchuna, according to records, was encountered in the 1950's. This is not actually very long ago in the concept of primitive entities fleeing and finally dying out before the drive of civilisation and of the opening up of new communications and technologies. The Siberian Arctic is still immense and desolate enough to harbour a few scattered wild types, even if they are officially supposed to have died out.

Pushkarev says that even in this decade, true stories of the Chuchuna are told in Eastern Yakutia.

The people said that until comparatively recent years, a wild man of the Chuchuna breed was living in the mountains of Verkhoyansk. He was more than two metres tall, covered himself in deerskin, and ate raw meat. He lived in caves in the higher altitudes, which suggests affinity with the Yeti's or Bigfoot's social customs. He slept through the winter. In the summer he was often met at dawn or sunset by herds men or hunters. On seeing a man, the Chuchuna generally fled, but at times he would utter a whistle and attack ferociously.

Local people said that sometimes a Chuchuna approached human dwellings. If the man of the house was not there, he would break his way into the homestead and carry off women and children. He would take the woman as a wife, and two or three years later a child would be found wandering, gone wild, and having lost the ability to speak or reason.

The younger men of the herdsmen communities would be sceptical and consider Chuchuna talk as tales told to them in childhood by grandparents. A valid point they did make to explain the tales was that long ago the elders would frighten children with exaggerated stories of horrible wild men simply to stop them from venturing deep into the forest where the real wild animals could be much more dangerous.

Pushkarev and Pupko encouraged the persons they interviewed to give their own versions of the question without interrupting with their own interrogations during the course of their descriptions. This was more productive of straightforward accounts. These came mostly from people whose parents had seen the Chuchuna. Some reports went back to the 1920's, and sometimes even to pre-revolutionary days.

Here is one story they heard from a 55-year-old woman. They took it down in writing, sitting on the banks of the river Khoboyotta among a group of reindeer herdsmen. She was an Evenki by race, the Evenki being one of the native tribes of Siberia, and she lived in a village, Suordakh in the Verkhoyansk region. She told how after the Revolution some people of her village met a Chuchuna once when they were out gathering berries. The Chuchuna too was picking the berries, stuffing them into his mouth, and when he saw the people, he stood up very erect. He was extremely tall and thin, more than 2 metres high, said she. His arms were abnormally long reaching right down to below the knees. He was covered in deerskin and was barefoot. He had long matted hair, a big and broad chin, much bigger than a man's, and a black face with a forehead that jutted out like the peak of a cap, she described. He was very like a man, but very much bigger. After a second, he ran away very fast, and he gave a high leap at every third step.

Pushkarev and his friend took down several such clear accounts as this one, and the same Chuchuna stories tallied in detail, whether the tellers were Evenki or Yakute.

Elena Afanas'evna Gorokhova too had a Chuchuna story. This was a direct personal experience of the early 1920's. In her own words: "Soon after the Revolution in 1922, we were mowing hay in the valley of the river Hiya" a tributary of the Yana - "Round about mid-day I was cooking dinner in a glade not far from the stream, the men still away haymaking, and I was along. Suddenly about 40 metres away in the bushes I heard a crash, and into the glade came a most unusual man of huge height, three head taller than our menfolk. He had long tousled black hair, his body and legs had deerskin all over, and his face was very dark. I could not make out his features for he stooped a lot, and his arms were so long they reached right down to the knees.

His body was all covered with his own shaggy fur. He rushed across the clearing not even glancing at me. Probably he did not see me. I was terrified, for I had no idea what it was. In the evening when the party were all together, they told me it was a Chuchuna."

The two researchers did not confine their investigations to the places mentioned but carried out enquiries in the Namsk region of Central Yakutia, and they talked to people who had come from Vilyua, but in neither of those areas did anyone know anything about the Chuchuna. Nor was anything known about them in the dwellers in the northern part of the Amur region.

What surprised them was the narrowly circumscribed area of Chichuna encounters. Their search led them even further and further eastwards, to the northern regions of Kilyma and Chukotka. In the central area and the western reaches of Yakutia, the Chuchuna is either totally unknown, or if he is known, one is referred to the distant land around the Verkhoyansk Range.

In the North-East of Yakutia, in individual places, the Chuchuna was frequently encountered, while in others, the phenomenon, if it can be so called, appeared rarely.

It would seem that this was a distribution in one comparatively small area of some unaccountable natural reason of food availability, perhaps, or of safe quarters for concealment from humans.

In the 1950's only two encounters were registered, both on the river Adych. The valley of that river, a Yana tributary, is one of the most inaccessible parts of the whole of Yakutia. Even in the 1970's, men are rare visitors unless they can travel 800 kilometres along a savage river full of rapids, between steep rocks and swampy taiga forests. Otherwise, the traveller would have to go there by helicopter. The forebears of present-day reindeer herdsmen used to see the young of the Chuchuna swimming across the river at the spring thaw of the ice and jumping out to steal food from campers.

Pushkarev and his fellow searcher, found one fact more puzzling than all the puzzling facts about the Chuchuna, and this was the limited period of time during which sightings were recorded. According to all accounts, the Chuchuna were seen around at the end of the 19th century and the beginning of the 20th. He was fierce and powerful and on confronting a man was often the aggressor. One report of a Chuchuna who had been killed, described how when the tight deerskin was peeled off the body, they saw that the whole of it was covered with thick hair, or fur. There are many accounts of confrontations between the Chuchuna and local hunters, and generally there would be a killing either on one side or the other.

A well-known Soviet historian and ethnographer, Gavriil Vasil'evich Ksenofontov[4], wrote of the Chuchuna: "...he is a man. He lives by hunting wild deer. He devours the

[4] https://peoplepill.com/people/gavriil-ksenofontov

flesh raw. He lives in a burrow like a bear. The Chuchuna wanders around solitary, and at sight of a man he runs away, or shoots. It seems he had a wooden bow..."

A wooden bow? Where had they known how to make one? They could shoot to kill, and yet did not know of the use of fire. They possessed human characteristics and animal ones too.

The Chuchuna of Arctic Siberia were obviously no legend, though they may have been useful bogies to frighten children. Though they migrated from somewhere in the East, none could tell from where exactly.

Some of their physical traits equated with those of the Snowman, Yeti, Bigfoot, Almasty, Sasquatch. He is a part-removed cousin, and if there is any remnant of him in the unvisited wilder parts of Arctic Siberia. An avant-garde scientist, if such a category exists, might even call him one of the missing links.

There is one certainty. Which is that he is a fellow traveller with my previous examples.

From those far-away records, here is a leaf down the decades to more recent times. Also, thanks to geologist Vladimir Pushkarev. He described this modern incident in one of his published reports which I obtained thanks to the Darwin Museum, Moscow, and which was recently (1983) quoted in an interview I gave to the "Sunday Times" of London:

"Two country lads were mowing grass in the Caucasus Mountains only 500 yards from the Nalchik Pyatigorsk Highway, often used by tourists. They happened to look up. By a water duct running through the upper meadows was a young hair or fur-covered creature looking down at them. They watched, amazed as the figure entertained them with jolly antics, running with great speed to and fro on the duct edge, jumping from it, somersaulting and putting on a fun act for their benefit.

He seemed to want them to come up and join in the game. The two human boys, not scared, but confused, stood there rooted, and the poor disappointed 'homi' (one of the many names the Russians give their mystery creature) vanished in a hemp plantation. When the boys told their mothers, the mums said: "Oh that was only a young Almasty".

From an informant in the Himalayan area a more recent report reached me of sightings, quite independent of the Yeti publicity often encountered in these regions.

Local Sherpas and villagers said that distant figures seen on high slopes had been noticed and were very visible. They were big, fur-covered and hunched, looking bear like, yet different from bears. They had appeared before. Local people said they lived mostly by killing cattle. I do not know if they were vegetarian as well as carnivorous, the same as the Canadian and American version is said to be. They only attack humans if closely confronted. They go for the head. The Indian villagers thought that the creature always attacked the headfirst. This ties up with the belief in several countries that the "animal" wants to get at the brains a delicacy for them!

An important local belief is that the creature is half-animal, half-human. This is a belief expressed for many years by students and researchers of the Yeti mystery in many lands, including facts by the late Professor Boris Porshnev.

Chapter Fourteen
The Yowie Down-Under

Where mysteries of nature are concerned the hunting season is elastic, depending more upon climate and opportunities than the classic dates and areas of sportsmanship.

Phenomena of nature can be investigated at all seasons subject to the mystery, generally a living one, finding suitable weather conditions and food supplies.

Tony Healy of Canberra, Australia, called upon me following a finding-out session at the Canberra National Library, and apparently after talks with Graham C. Joyner, another mystery searcher. Graham C. Joyner[1] has written a documentary on Australia's Hairy mystery, of which there will be more further on. Tony Healey also met Tim Dinsdale[2] of Loch Ness fame, an author too of several books on his subject, which I refuse to call a monster because it and it's like have been known up and down the lochs and coasts of Scotland and the Western Isles for generations.

Tony was adventuring, not only at Loch Ness, but in the Irish Loughs, and would later go to the Americas. He is in search of the fabulous water-creatures of vast size of many lands, and before returning to Canberra will pay a second visit to Lake Okanagan in British Columbia, also reputed to harbour a similar water creature. Tony was in Vancouver in 1978 at the British Columbia University conference on the Sasquatch where I too attended. He told me he had intended to see me while there, but somehow, we missed each other as can happen during a very full and continuous programme.

As Tim Dinsdale had introduced him, this was good enough for me, and Tony was very welcome, even if his special adventuring was after a too watery fellow-traveller for me to include it in this collection. But being interested in the Yeti search too, he brought me news in detail of its equivalent sphere in Australia.

There have been only very sparse rumours in the West about the Australian Yowie, or Yahoo as Aborigines used to call this strange ambulating entity whose unexpected "walk-abouts" were often bogies to scare children not to wander too far in the bush from the family camp fires.

[1] https://www.abebooks.co.uk/9780908127009/hairy-man-south-eastern-Australia-0908127006/plp
[2] https://en.wikipedia.org/wiki/Tim_Dinsdale

In recent years it has been treated with the levity that was, and still is at times, accorded to the Yeti, or Abominable Snowman of many names. But currently the Yowie is assuming substantial form. Researchers have come out more in the open in the Antipodes, and though the Yowie is still considered very fringe natural history in the popular everyday mentality, more serious thought has been applied to the question beyond jokes and dismissal as mere superstition and Aborigine fairy tales.

In quests of this nature, it should be remembered that reactions to anything hitherto unproved or strange are always the same the world over. Let us have our folklore and fun by all means, but let it stay there. And it does occur to one without too much malice that even the most honest and genuine of scientists may dread discoveries that might necessitate re-writing laborious texts.

Tony Healy planned to end his water-creatures world tour with a trip to Nepal where he hoped to track down some more Yeti questions, if not a Yeti actually, so adding one more adventurer on that trail of many years.

Having followed the Yeti saga, myself for several years too, hearing about such a remote fellow-traveller as the Yowie of Australia was refreshing and original news.

The Yowie distribution seems to have been mostly in the Southeast of the continent veering to North-East, and several encounters took place in coastal forest areas. Among many new reports was one from Queanbeyan. This referred to huge humanoid-like footprints, and one observer said they could only have been made by a creature of giant proportions.

This was also the opinion of Mr. Rex Gilroy. He is a field naturalist well known for his twenty years of Yowie research, even though only thirty-four years of age. He is recognised as an Australian authority on the Yowie, also known to the Aborigines as the "Great Hairy Man". **[Rex Gilroy died in 2023 whilst this book was being prepared for publication Ed.]**

Scepticism has arisen nonetheless because this "Great Hairy Man" figures in Aboriginal folklore, and folklore can confuse issues, even when physical evidence of these extraordinary footprints indicates that something alive and very heavy must have made them. Especially as they have occurred in wild territory where hoaxers would certainly not take such fatiguing and non-productive trouble to do any faking just for a joke.

One set of footprints was discovered by a motorist who pulled up for lunch on a deserted scrub-covered roadside some miles out of Queanbeyan, according to Mr. Gilroy. They were later found to be exactly like another set of large footprints near Tidbinbilla. Both sets measured 16 inches in length and were more than a foot wide. This makes them considerably larger than the historic footprint Eric Shipton photographed on the Menlung Glacier in the Himalaya in 1951. These were about 12 ½ inches by 6 or 6 ½ inches. The Yowie tracks appear to equate in size with those of the Bigfoot or Sasquatch of the Americas.

Footprints of similar proportions to the Queanbeyan ones were discovered during April 1977 on the rugged Carrai Plateau west of Kempsey on the New South Wales north coast where Rex Gilroy had led an expedition to several north coastal districts to find further evidence of the Yowie.

Gilroy said: "Casts made from these footprints compare favourably with those found in the Himalaya which are thought to have been made by the legendary Abominable Snowman or Yeti." Gilroy thought this might imply that the Yowie may be even more closely related to the Himalayan Yeti than to its other cousin, Bigfoot of Canada and North America.

This could be possible except for the discrepancy in footprint dimensions, though here the only yardstick is the footprint in the Shipton photograph. Round about that time there were others taken, but none as valuable evidence as Shipton's, though the tracks photograph taken on Annapurna by Don Whillans in 1970 runs it a close second.

Gilroy calculates the approximate height of the creature that left the Queanbeyan, Tidbinbilla, and Carrai Plateau footprints to be twice the height of an average human being. "This would place the beast at over ten feet tall," he is reported to have stated, with the usual understandable ambivalence of persons who refer to the phenomenon at one time as an animal and at the next as near-human.

In 1978 Gilroy established the "Australian Yowie Research Centre" at Kedumba Emporium, Echo Point Road, Katoomba, New South Wales. Leaving nothing to chance in such a vast continent, Gilroy quotes his telephone numbers as (047)-82-2376. He asks for any persons able to help his investigations with news of Yowie sightings or any other evidence to write to them.

Soon after opening the research station, he investigated reports of Yowie sightings on the New South Wales southern highlands around Bowrai and Mittagong. A month before, a farmer near Bowrai tending cows in his paddock which was on the verge of thick mountain scrub, sighted from a distance of only one hundred yards, a taller-than-man sized hairy humanoid looking beast as it strode bipedally from out of dense bushes, and stood there staring at him from over a fence.

Another report, recent at the time, told of a motorist on the Mount Kosciusko Road. He was surprised one night when a seven to eight feet tall apelike, yet manlike form ran across the road ahead of his car. The creature was dazzled for a second by the headlights and disappeared into the roadside scrub.

Gilroy has amassed a huge collection of reports of sightings and other evidence, doing for the Australian phenomenon very much what John Green has done for the Sasquatch of British Columbia. Gilroy has an enormous amount of data from over a wide area of Australia, but chiefly from the eastern part of the continent, as well as from Tasmania. He said how in New Guinea the native inhabitants speak of a terrifying hairy monster they call the Kilboornee, or Hairy Devil Man.

Which reminds me of the Yeti-type creature of Borneo across the seas to the North-West of New Guinea, called the Beruang Rambai. Somewhere I still have a letter from a gentleman who lived in a longhouse, who was delighted, I was told, that someone was interested in their type of Yeti!

Unfortunately, the letter was in the Dyak tongue, and I have never met anyone who could translate it though I know its gist!

Yowie reports in Australia seem to have been coming in large numbers of recent times. Possibly because any of the public who have had experiences, are not so fearful now of being laughed at. This is probably still the reason for some informants wishing to remain anonymous. Like the man from Narooma. Numerous similar reports come from other inhabitants. However, this man told of a strange bear-like creature. He saw it on two separate occasions. Then he saw three in one place, the next time he spotted one, and then two more. They were completely covered with hair, he said, with head set straight into the shoulders with no indication of any neck. The fur or hair was rusty brown, and they walked with a stoop. The man got near enough to see them facially. He described how their eyes were "big, beautiful, and doe-like". The description reminds me of Merlin Hellener's childhood encounter in the Cascade Mountains of America, and her impression of the mystery invader of the little girls' bunkhouse looking down at them with intelligent curiosity.

The Australian Yowie seems to be a fellow traveller. Many characteristics are similar to those of the many-named Yeti, as shown in another account from the anonymous man. He and a friend were out hunting fox in the same area. They saw no Yowie but heard crashing about in nearby heavy scrub, and they smelled a typical musty smell associated with most reports of the phenomenon.

Gilroy told of an encounter he once had with what he thinks must have been a half-grown Yowie. He was having a sandwich at a very deserted wild area, at a rock known as Ruined Castle. He had a feeling of something watching him, when suddenly a small apelike shape, like a small man, broke cover and shot across the track to the far side where it disappeared in heavy undergrowth. Gilroy says there have been more than 3000 sightings of Yowie since white settlers came to eastern Australia.

One was reported near Sydney Cove in 1795. This was the first report. But long before that the Yowie under other names had been established in Aborigine legends. Later sightings occurred in the 1850's in the blue Mountains. In recent 1978 skiers said they saw a tall apelike animal near Mount Kosciusko in the same region where such creatures had been seen before. A farmer of the Southern High lands in 1979 told of a Yowie-type creature vanishing in the scrub near his holdings.

In Bill Harney's delightful tales and poems of the Aborigines with whom he lived and made his life, there are many references to the Hairy Man or supernatural entity; the strange spirits with even stranger names that seem to alternate from being benign influences to becoming creators of evil happenings. They have long, evocative and difficult titles. I shall quote only one, the Wulgaru. Sometimes he is a good Devil-Devil, sometimes a mischievous one. It seems to depend on the mood of the moment, and the Wulgaru must command respect.

Which is not always what serious investigators receive when in the course of their efforts they have to contend with disbelief. But searchers after the truth continue to search and are rarely deterred.

Graham C. Joyner mentioned previously has produced painstaking direct data in his short but fully informative book, "The Hairy Man (of South-Eastern Australia)". He gives all the Yowie or Yahoo names, and difference Aborigine languages. He quotes authentic cases from individual witnesses. Most of these, as well as newspaper reports, have been lodged for years in local and national libraries.

His presentation is a very good, condensed documentary without bias.

A quote from his foreword to his book makes a fitting last comment for this chapter: "... to many of the questions which might be asked about the hairy man there are, as far as I know, no answers. Perhaps it is, like the tale of the Giant Rat of Sumatra, a story for which the world is not yet prepared."

Chapter Fifteen
Mystery Gaps

In recent years some controversy arose over Darwin's alleged missing fossils in his research. One could find an indirect link there with the somewhat similar lack of fossil proof in the Yeti's disputed existence.

Some scientific spheres were questioning the hitherto acceptance of Darwin's discoveries, or lack of them in one particular quarter. Contenders pointed out certain blanks in Darwin's theory of evolution. Such as his lack of evidential fossils at certain times in the world's history. Critics did not dispute his evolution theory, but considered he did not find links in fossil evidence to maintain his claim of natural selection... survival of the fittest. Apparently, Darwin had given no reason at the time for certain lack of fossil evidence.

Among some of the challenges I have turned up a paper by Francis Hitchings, which appeared in the October 1982 "Reader's Digest" He quotes quite correctly the evolutionary process expounded by Darwin and others:

1. Bacteria and slime
2. Sponges and jellyfish
3. Fish with backbones
4. Amphibians living partly on land
5. Reptiles, including dinosaurs
6. Birds and mammals[*]

Darwin considered the six stages were linked and happened in a bit-by-bit process of improvement. Result his Natural Selection. The survivors of what we might call nature's experiment in growth. It should show down all the ages. But it does not, because of the several gaps in fossil discovery evidence. At different stages of the earth's development such gaps are evident... fossil links between important groups of once living creatures are absent at certain times whereas there should have been a continuous chain bearing witness to the march of nature.

None-the-less I believe that Darwin did not make such a mistake, even bearing in mind the frequent gaps in fossil proof, even if at the time he gave no reason for this lack.

---

[*] RF - Birds are in fact dinosaurs and ergo feathered reptiles.

Such reasons were... and they exist to this day... the geographical and climatic unpredictable factors in earth's history. Changes in nature have been unexpected and have occurred during the millions of years of this planet's existence. Gaps in fossil proof could have been caused by conditions of change such as those that destroyed skeletal evidence of the Yeti's existence.

Nature is untidy in its climates and cataclysms. Dinosaurs died almost suddenly because of this. Animals once crossing back and forth by land bridges from Africa to Europe were caught by floods and vast upheavals. Only a few exotic bones were found hundreds of years later in some of Europe's caves.

Perhaps if Darwin had lived longer, he might have given those vagaries of nature the reason for those disputed fossil gaps.

When faced with a debatable mystery of the once living or still living world some scientists and "science-tasters" repeatedly return to their conviction that only physical remnants of a once living creature is proof of its having once existed. They call for this proof when in many cases proof has been destroyed by predators, climate, or earthquakes.

The late Professor Boris Porshnev when challenged to produce bones to prove the existence of the Yeti/Snowman/Almasty etc., once replies "The Snowman and the Snow Leopard live in inaccessible mountains of our remotest regions, which at times are battered by heavy storms and overflowing torrential rivers and waterfalls. Neither bones of Yeti or leopard are found because the forces of nature break up and destroy them."

Such are the conditions that govern development and descent. The Elephant, as once described by Dr. Malcolm Coe[1], dies in the bush. Generally, in a short time, the predators, small insects or beasts of stature, have done the cleaning up and little trace remains.

It is from these causes that we get the phenomena and the alleged mysteries. They are found in the imponderables of the world.

Yet the imponderables are sometimes confounded by reports of almost unbelievable physical factors, and alleged discoveries of hairy primitive giants, difficult to trace

---

[1] https://prabook.com/web/malcolm_james.coe/537108

because of their wild and remote habitat, yet that nonetheless possess some aspects of truth.

The American Indians could probably tell outsiders much more than they have disclosed already about their ancestors' memories of giant, yeti-types associated with the remotest and most inaccessible mountains. Some of the Indian tribes still talk of those tales at times, but so far, they relate also to folklore, so were often not taken seriously. But their stories handed down by forefathers are listened to with more understanding and logic now, for in current times the Indians are more modern and up-to-date, and education has widened nature studies and horizons that are making history.

Chapter Sixteen

In a book of this nature the writer has to play it by ear concerning place and presentation. Because very often new data is discovered at unexpected times, and even when the last chapter is written; Such is the case in this book. New material here may simplify my term, imponderable, used earlier in one chapter. Just as in my final chapter I shall quote an experimental seance exactly as it happened, followed by its strange sequel; so, I shall record my newest data.

My informant is Art Kapa who with colleagues runs the Bigfoot Investigation Centre on the edge of dense Michigan forests in America. In part of the American press interested in the natural sciences some of my work has appeared. Kapa wrote to me, and we are in regular research touch. His group's experiences in the Bigfoot mystery make their long search and evidence more believable than many of the scattered reports that reach me.

I asked Kapa why it was that the authorities did nothing to solve this long-standing distribution of hominid-like beings in the wilder and most remote regions of the country. Isolated forests, swamps and mountains do have official forest rangers visiting the various wildernesses. Why is there seldom any public news about them? Kapa replied and I will later quote him.

For many years inhabitants on the edge of Michigan forests have known of Hairy Men, wild creatures known in different continents, as Yeti, Abominable Snowman/Bigfoot/Sasquatch, and other names, including the Russian titles for the phenomenon.

I cooperate in this fringe science. My experience now suggests that it should not be "fringe" for much longer if only the honest truth were applied.

"Uncle Harry" is the humorous term which Art Kapa's group have applied to the Hair man-beast, or beast-man of their terrain. This community are keen woodlanders working in timber and occasional grain cultivation. Identifying and collating records of "Uncle Harry" is their quite important side-line. The hairy creatures of the almost impenetrable forests have proved harmless (as is known to experts) except if they feel endangered. They are curious about human beings and prowl around human homes at nights in their inquisitive search, and for foraging dustbins.

Art Kapa described how one evening there were other similar cases... the local dogs set up wild barking. Grunts and howls were heard coming from the deeper woodlands. One dog was so crazy to get loose that he bit through his kennel chain. Kapa and researcher neighbour Ray Atwood investigated but did not dare penetrate too deep in the woods. Next time, Ray was cutting firewood in daylight. It was a nice day, warm with fitful sunshine. As Atwood ran his chainsaw he heard a chicken squawking in the woods. The sound came from between where he stood and a neighbour's holding on more open ground. This neighbour had lost a lot of chickens in the last two weeks, and Ray found; whatever was there moved very fast... faster than a man would walk, and Ray hurried to keep up towards a small opening in the woods. He wondered if there was a stray dog or fox about. He walked towards them and stopped to watch.

Out of the woods walked a very large creature. On two legs and looked human but seemed 8 or 9 feet tall and covered with thick brown hair that shone in occasional sunbeams.

As it crossed the small opening the Bigfoot creature was hanging on to the chicken which could still be heard squawking. The bigfoot's outstretched holding arms looked very long. Bigfoot experts, whatever the creatures' name or clime, always stress how long those arms are, reaching to the knees, whether their habitation is the Himalaya, Russia, or the Americas.

Before the creature vanished, Ray noticed the huge back, and then he really took fright and left the wood's edge.

Later Atwood came back with Art Kapa. They searched for tracks in the ground, but fallen timber was too thick to discover anything, though in a stream bed they found a footprint of about 15 to 18 inches long. But it was too difficult to take photographs and had been washed away next morning.

Atwood said that encounter was his life's most scaring experience and spoke of screams in the night from the woods, waking him from sleep. Atwood had built his house half a mile from the county road on the wood's edge. He said the Bigfoot he encountered had a very bad smell. Often described by other investigators.

Kapa gave me reports by other persons. Spots were found where some living very large creature had moved, and a neighbour, Wayne Legue, said that recently he heard a Bigfoot steal another chicken. Kapa said "I have been sitting at night with a live chicken in a cage, hoping for a sighting of the wild hairy being, but so far.... No luck. I will keep going back every night. With luck I may get Mr. Bigfoot to come to me."

Following is the revealing letter I received from Art Kapa in reply to my questions.

"No one clearly knows the truth re the authorities' attitude. Most of them have closed minds. They will not even listen to the ideas of people with attitudes different from their own. The ones who do believe in Bigfoot won't talk about it for fear of ridicule and for the stability of their jobs.

"The officials feel that if Bigfoot is not caught or killed where they can inspect it, then the thing just doesn't exist. Since it doesn't exist, in their minds, then they feel there is no reason for cooperation.

"As far as the government goes, it is hard to say if they know anything. We know that they have something in a hangar at Wright Patterson Air Museum in Ohio. It is heavily guarded, and no one is allowed near it. Our government keeps a lot of things from us. I figure it is because they can't explain them. If, let's say, they have an alien being or a Bigfoot in that hangar the government would never admit to it. For years they've been saying that U.F.O sightings have been from marsh gas or a star that came out of its galaxy into ours. It would be embarrassing for them to have to concede that they have been wrong all these years. With a Bigfoot if they find it is close to human, they may even have to rewrite Darwin's evolution.

"Eating crow is hard enough for anyone, but it is especially hard for officials. As far as 'ignoring us common sense country folk' as you put it, that's exactly why. Because we are country folk without the sophistication they have in their education. According to them we are just trying to get into the limelight without working for it."

Rumours went around that Bigfoot had been shot here in Michigan and it was sent to the State University in Lansing, the state capital. But needless to say, no one knows for sure. If it did happen everything is again being kept hush-hush.

"As far as I am concerned, it makes no difference how many scientific degrees you have, if you don't have common sense the degrees aren't worth the paper they are written on. It takes a person with common sense to reason out all the evidence, such as: footprints and actual sightings, throughout the U.S. and the rest of the world.

"There is a manlike creature roaming this earth and has been since prehistoric times. The wilderness area in this country is so vast that regardless of the population

145

explosion, we have hardly made a dent," said Kapa. "Most reasons a Bigfoot is sighted is purely curiosity on Bigfoot's part. Curiosity sightings outweigh accidental sightings, such as: Bigfoot watching a woman picking vegetables in her garden, loggers cutting trees, a farmer working his fields, children playing, or just Bigfoot looking in the windows of a person's home. Because of the unusual sounds of a T.V., the lights in a house, or the curiosity of us alone; since we are almost identical to them, except for size.

"All animals are curious, but none as curious as Bigfoot is about humans. I believe Bigfoot is our closest living prehistoric humanoid.

"It's a shame one will have to be killed in order to prove they exist, or worse yet, to capture one and put it in a cage would be an inhumane injustice.

"To be able to see one for myself, and photograph it, would be a reward to me above all other rewards. I have no fear of harm coming to me, from a Bigfoot, but I don't trust them 100%. If we were to surprise a family group, it is possible the male would consider me a threat to his mate or offspring. If a man acted with some degree of common sense, he may be able to get himself out of a predicament like that. Perhaps someday I will find out hopefully.

"The only person I recognize, and have respect for, as an authority on Bigfoot is René Dahinden. Most of the rest of investigators are gatherers of paper reports and enjoy the limelight of the media. Those 'armchair' investigators do nothing but give the honest investigators a bad name. Sometimes I wonder why I investigate, the cost greater than I can afford. But it is complete fascination of something that isn't supposed to exist but does. The idea that this creature could have been the starting of our race keeps me going. I want to know for myself, I don't give a damn what authorities think, exactly what Bigfoot is, human or animal.

"I hope I have answered all your questions. If you ever have a chance to come to U.S. you will be more than welcome to spend some time here with my wife and myself. I would enjoy sitting and talking to you in person. Sorry I took so long to answer. I was busy clearing snow and keeping deer fed. Sincerely.... Art Kapa."

Back to my views, Art Kapa and his ilk are genuine humans of great sense and feelings. This world mystery may still remain imponderable, however much new evidence increases, because of a turbulent world striving for the wrong progress instead of the humanities. Governmental non-action is fear of long-accepted scientific facts causing world embarrassment. They know something but will keep it secret as long as they dare.

But something super-normal may happen, and the truth about the long hidden wild Hairy Man will be revealed.

Even as I was finishing the last sentence about Art Kapa's Michigan Bigfoot experiences, I received a follow-up of his research in the hills and mountains of the adjoining state of Indiana, and as he set it down for my benefit here are his words again:

"My trips into Indiana brought me more cast footprints than pictures. It seems our quarry is very shy of cameras. A few sightings took place as I was travelling around in my search. One man contacted me with the news that he saw one of the creatures crossing a river. He found 18-inch footprints on the riverbank and nearby sand bars."

Art Kapa tells me of a neighbour who saw one of the Bigfoot type poking around his backyard at 11.30 at night. Footprints this time were 14 inches long. They vary, probably with age. Safe equipment is very necessary, he says, and he is acquiring snake-proof clothes and boots as the deadly cooper-head abounds in the hills and caves. He tells me that he will investigate those caves in hope of finding evidential bones or other proving objects. He will explore again in the fall (autumn) as then the underground water reduces to a safe level. Bigfoot must go somewhere to die, and there is a record to suggest that they bury their dead. While he was on the Indiana trail, a grey-coloured Bigfoot was seen which would imply age rather than natural colouring. Bigfoot hair or fur colours have been described for centuries as varied, as occurs with the human race.

Kapa is going back to the Michigan forest borders, as that is where he works his timber and vegetable holding. So far (1986) there have been fewer sightings or sounds indicative of Bigfoot presence, but they are there.

It is a problem of collecting the right equipment and raising finances to cope for expert researchers as Kapa and colleagues. But he thinks that someday, some time, he and his helpers will be at the right place, and useful progress will get a chance to succeed.

All unfinished evidence of Bigfoot existence has been existing for centuries in the relevant lands concerned. Truth? When?

There are higher powers shaping human mysteries than human government.

There is more than body to every breathing, living shape.

Chapter Seventeen
The Silence

Foregone chapters attempt to present the wide panorama of unsolved Wild Hairy Man (Yeti etc) phenomena and various attributes. The species, whatever it is, varies in certain features according to geographic distribution. But they all indicate a relationship. As in the known human races there are differences in dimensions, colour, and reaction to surroundings. Some differences in the Yeti/Snowman etc, images are very marked, but they come from the same roots.

The late professor Boris Porshnev used arguments that tally considerably with up-to-date questions and present time theories. His views widened as he and followers studied, experimented, and travelled. And the remarkable group who have inherited his teachings, the Darwin Museum in Moscow, continue his work. This work has in recent years widened considerably in cooperation with the International Society of Cryptozoology, and the wide experiences and opinions this Society provides.

But some perplexing characteristics in Snowman, Yeti, Sasquatch and Bigfoot findings exist today as they did during Porshnev's career. An important question to solve is the alleged double nature, both animal and human, of the problem. This could be one of the greatest problems to deciding the truth. The situation has operated exactly the same since the long past days of Imperial Russia and the indifference of that nation and the West when Colonel Przhevalsky's discovery of the Almasty (whom some students now think are extinct) was pushed under the carpet, as an embarrassment to bureaucrats, down to the present world-wide avoidance of the phenomenon subject.

Much study and conjecture about the Missing Link took place down the years. Yet Homo Sapiens may have had its missing link on the tracks of the wild places of the world and looked the other way. In view of the increasing reports of strange, hairy and primal ape-like creatures encountered in almost inaccessible wilds which scientists and hunters penetrate more often these days, there are clues to unravel. This occurs at times quite often in remote mountains and forests of China and the Mongolian heights. Perhaps the ultimate discovery will be there.

Years ago, I devised and designed an S-Map based on my research of likely distribution of the Snowman trail in all its many names. The design formed a large S.

lying on its back, and covered all the areas of this alleged distribution, from the Caucasus Mountains to northern California.

Sense of superiority is death to human understanding. Man dares not think that he, lord of creation, might have dropped some untidy physical mystery in the millions of years of planet earth's existence.

But nature is untidy, sometimes wasteful, and very mysterious. Some of nature's most mysterious happenings was the migration of primitive Asian tribes across the Bering Sea land bridge of the American continent from the Russian vastness, thousands of years before the melting of the ice cap there. Some sub-human races might have trekked from Asiatic land masses, from Siberia's Arctic regions, and thus created the puzzle of Bigfoot/Sasquatch hairy bipedal creatures. This vast gradual movement might even have carried over some of the Arctic Siberia Chuchuna species with their puzzling human characteristics that yet had a most confusing animal mixture to their physique and customs. All those mysterious groups were possibly motley refugees escaping at different epochs from famine, plague, or land disasters.

However, the Russians of former days of alleged civilization were much like their opposite numbers in later centuries, and preferred folklore to fact, tending to relegate the Yeti/Kiik-adam/Almasty/ Snowman riddle to fairy tales.

Looking back to the favourite argument against the Yeti that no bones or remains are ever found, one can compare the life and death of known species in remote wildernesses. Body or skeletal remains are seldom found, and climate or predators are the reason. This was not only Professor Porshnev's views but are beliefs of contemporary scientists, such as Dr. Malcolm Coe, of the Animal Ecology Group, Oxford. He once kindly answered my questions. He wrote: "My manuscript on the elephant carcase is only just being completed so it will not be published for some time. The brief picture is as follows: 1. Soft tissue disappears in about two weeks. 2. Skin disappears in three weeks to six months, depending on the season. 3. Bones are scattered by scavengers and elephants, but the larger ones will remain visible for at least ten years if they are not buried. The rate clearly depends upon climate."

These facts refer to animals that exist in large numbers. How much more difficult to find traces of mystery creatures of the rarity of the Snowman or its fellow travellers.

The many hoaxes carried out in ignorance have been detrimental to serious research among scientists up till recent years, but records do show that several experts are now of the opinion that there is something factual to discover and they do not dismiss the subject as an invention.

When I was in British Columbia a busload of travellers near a place called Mission saw their bus driver hop out in the roadway to face a large hairy creature ambling out of the woods, but the vision turned out to be one of a group of youths who had borrowed a large ape suit and were just having a spot of fun. The bus driver was scolded for deserting his bus!

I was slightly disappointed at the tame ending. It reminded me of a previous Canadian trip when two friends and I drove up the mountain loggers' track to a place called Harrison Hot Springs, high up the coastal range, an offshoot of the Rockies in British Columbia. A couple had driven up there some months before my visit, and the woman in the car called out that she had just seen a Sasquatch. Her companion emerged from the side bushes, but he did not see the Sasquatch which she said came rushing down from the higher peaks on left of car and rushing bipedally past the windscreen; he hurtled down several thousand feet to the shores of Lake Harrison below.

They told the story and then afraid of ridicule, left hurriedly for their home in Seattle, we were told. When anything unusual happens, the persons concerned take fright of being laughed at, and refuse to speak again. I was hoping something similar might happen when we visited the spot where the so-called Sasquatch had hurtled past that car. But no new Sasquatch appeared. It was bear country too, and even the sight of one of those would have been welcome. I could have been bundled into our car had one of those appeared. We could all have got in until Bruin had ambled past. I am not heroic, and as for the chaps with me, I don't think one is supposed to shoot a bear unless in dire danger. The greatest danger encountered were swarms of man-eating midges.

While I was revising this last chapter, I was trying to assemble a final assessment to the Yeti question. In spite of the good opinions on our research work, some colleagues and I wanted to organise some new experiment, and I wanted to create an unusual test.

It came to me in a flash. To organise a seance on the Yeti/ Sasquatch, Wild Hairy Man business. The still unsolved phenomenon is still so puzzling and contradictory in human response, that it has made me wonder all through this book if there is a higher reason why the world of men and nations seem to be everlastingly prevented from getting at the truth?

I am not hooked on the paranormal. But those of us who have open minds and study world questions and realities in any depth are well aware that "There are more things in heaven and earth....."

Yet in spite of increasing interest, research, and field work with undoubted evidence and clues, the Yeti phenomena, except for fleeting encounters, remain concealed in their forest and mountain wilds, almost totally undiscovered, and apparently protected from modern man.

For a reason?

A reliable clairvoyant/medium was invited, and she and a couple of interested friends and I sat at a table with the alphabet letters and the words "yes" and "no" in readiness, and the glass in the centre of our group. The clairvoyant lady said a few prompting words.

The following facts are exactly as they happened.

The glass slid about, seeming controlled by something. I asked if the Yeti/Hairy Man was alive. The glass spelt Yes/No/Yes. It might be because of the many names known apart from Yeti. At first, I mentally called whatever was sliding the glass 'The Glassperson', but now this force became very agitated and slid about rather dottily when asked if the Yeti/Hairy Man lived. The power spelt out a Yes, but after some aimless sliding it insisted that it should be called The Hairy Man insisting that he was a man and not an animal.

"Are we of the same family as the Yeti" was our next question.

The answer was affirmative, but there was great insistence that he was the Hairy Man, and not an animal.

Was the wild Hairy Man alive today, I asked, and he excitedly was, judging by the gyrations that went with a strong Yes.

Questioned about the Missing Link the entity replied that he was the Missing Link between Yeti and Man now!

This was the dramatic turn in the seance when the entity revealed his true personality... a living hairy man's soul projected to us from what forest or mountain wilderness we could not guess or even think about at such a juncture! It was humour, unbelief, toning down our reactions as the glass went sliding quite annoyed all over the table. Then the unseen but quite powerful informant told us that Yeti, and all were in Cuba. This was quite ridiculous, for Cuba in the West Indies was most unlikely terrain! To get on neutral ground I asked if governments were blocking Yeti news, and the answer was 'Yes'.

There were nice, kindly exchanges when Hairy Man seemed in better mood and there was quite a polite fade-out. We had not learnt much, and I was still sitting on the side-lines as far as seances were concerned.

There was a sequel two nights later at home when I was going to bed.

Midnight. Since the seance I had not had time to consult the Britannica or a geography book over the Cuba nonsense to try to elucidate what the Hairy Man had been driving at. Something suddenly shot across my mind. It was very urgent, a peremptory order. From a bookshelf I took down a world map gazetteer. Another mysterious mental streak made me turn to Page K of place names.

There it was at the top of the page: KUBA, (U.S.S.R) a spot at the tail-end of the Caucasus Mountains, near the Caspian Sea: The Caucasus, for centuries an area associated with reports and sightings of strange, wild half humans ... perhaps three-quarter humans.

I went to sleep saying thank-you. I know to whom. Yet do I?

We may still be only dimly aware of unexplained, unfinished intimations in nature because after the dawn of life we gradually alienated ourselves from creatures with whom we shared the earth.

In this book's pages I have included records of some unusual physical species, and how some of the freakish deviations in living souls may have been difference manifestations of life. And developments may relate to one another of those manifestations, even when indirectly.

Myth began in the long past from a grain of truth in an age forgotten moment in time. Though I have called the Yeti/Wild Hairy Man Anthropo-zoology, it might be that someday this name I have coined will be accepted as fact.

Once I though in earlier studies that the Yeti/Bigfoot/Sasquatch/Metch Kangmi was an unidentified anthropoid. Now I do not.

There is more than body to every living shape. Descent of man from early half-man would be no shame, for inarticulate awareness of the spirit behind all life appears to have been expressed even by stone-age ancients, as some of their sensitive customs showed. So called civilisation forgets. The unseeing mind cries out: "Here is something strange and not us. Let us kill it."

So, the world's silence remains. But the mystics know.

In William Butler Yeats' poem "The Second Coming" the last two lines makes no reference to the inadmissible Missing Link, but they are saying something:

"And what rough beast, its hour come round at last, Slouches towards Bethlehem to be born."

Odette Tchernine F.R.G.S
December 17th 1986

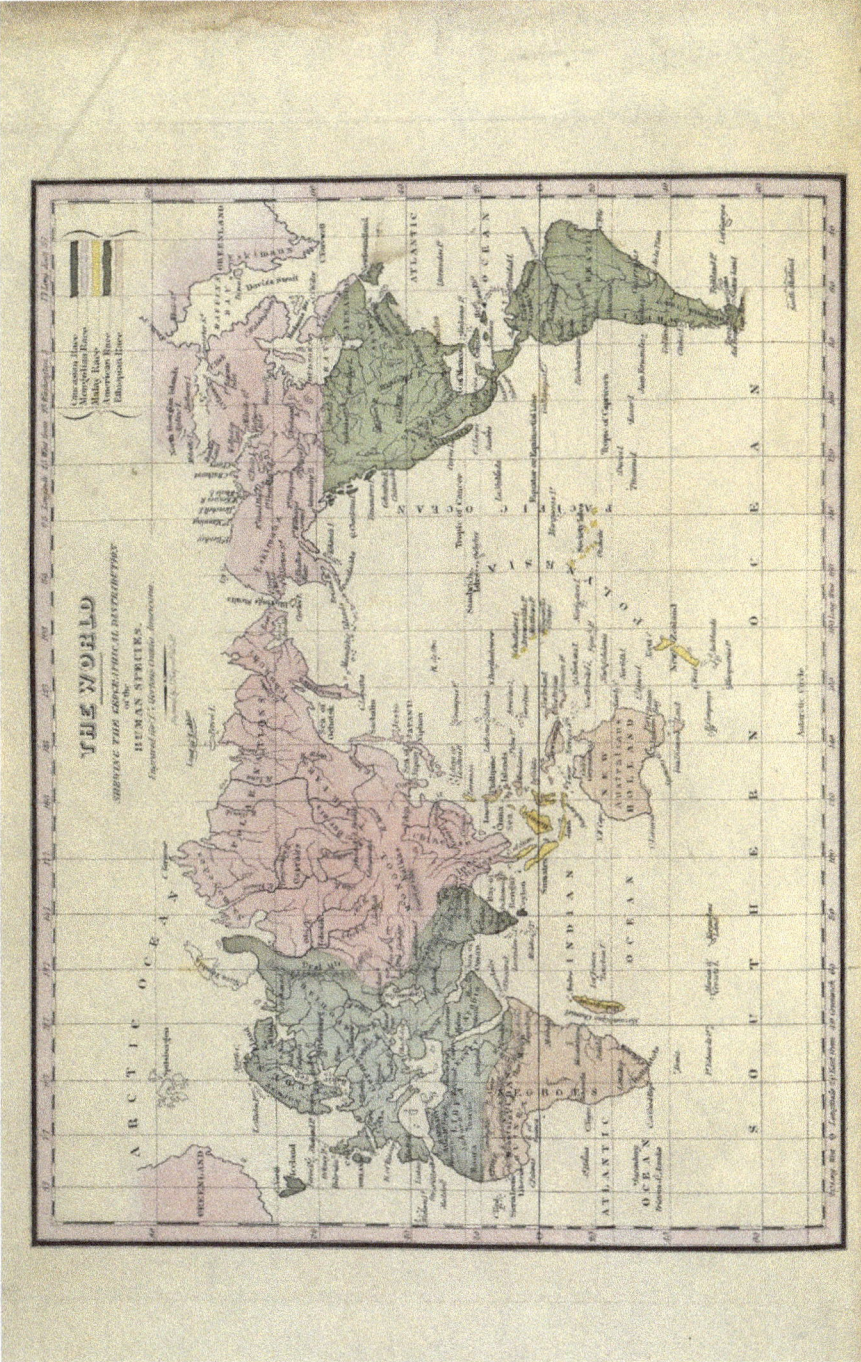

Odette's 'S Shaped map' will have covered from the Caucasus to Mongolia, a fringe of the Tibet-China borders, Siberia, and the Himalaya. Then over the Bering Straits through Alaska and down to British Columbia and Northern California. The original map is sadly lost.

Who or what is Bigfoot/Sasquatch or Yeti?

Bigfoot, also known as Sasquatch, is a legendary ape-like creature that is said to inhabit forests, mountains, and other remote wilderness areas, primarily in North America. It is described as a tall, hairy humanoid with large feet and a pronounced brow ridge. Despite numerous reported sightings and alleged evidence, such as footprints and hair samples, the existence of Bigfoot remains unproven and is widely considered to be a myth or folklore.

Yeti, on the other hand, is a legendary creature said to inhabit the forested, mountainous regions of the Himalayas in Nepal, Bhutan, and Tibet. Also known as the "Abominable Snowman," Yeti is described as a large, ape-like creature with black, brown or reddish fur, long arms, and a domed forehead. Like Bigfoot, the existence of the Yeti is widely considered to be a myth or folklore, and no conclusive evidence has been found to prove its existence.

While both creatures share some similarities in their appearance and reputation, they come from different cultural traditions and geographic locations. Despite numerous sightings and alleged evidence over the years, the existence of both Bigfoot and Yeti remains unproven, and they continue to be the subject of much speculation and debate.

The first recorded sighting of Bigfoot in modern times is generally considered to be the 1958 discovery of large footprints in Bluff Creek, California, by a man named Jerry Crew. However, there are many reports of similar creatures dating back hundreds of years among the indigenous peoples of the Pacific Northwest region of North America. These stories describe a variety of names and physical characteristics, but the general concept of a large, hairy, bipedal creature is consistent across cultures.

Below you will find descriptions for most of the different names used for Bigfoot or Yeti throughout Odette's book.

Abnauayu
The Abnauayu is the Russian analog of Bigfoot/Sasquatch, commonly linked to the Almas, said to roam the lands around the Caucasus Mountain range[1].

---

[1] https://cryptidz.fandom.com/wiki/Abnauayu

Almas/Almasty

Said to inhabit the Caucasus, Tian Shan and Pamir Mountains of Central Asia and the Altai Mountains of western Mongolia. The name is connected to a variety of place names (toponyms) in southwestern Mongolia, including Almasyn Dobo ('the Hills of Almases'), Almasyn Ulan Oula ('the Red Mountains of Almases') and ('the Red Rocks of Almases').

Folk belief in the almas in Oburkhangai and Bayankhongor has resulted in a name-avoidance taboo there, wherein the entities may be referred to as akhai, meaning 'uncle-brother'[2].

Beruang Rambai

The beruang rambai (Dayak: "long-haired bear") is a cryptid reported from Borneo, in both Indonesia and Malaysia.

Explorer Leonard Clark, who also encountered the milne and reported the Murung River bear, saw a bali djakai (Lawangan: "demon") at a water hole in Borneo's central mountains in the 1930's. According to George Eberhart, it "picked up a helmet left behind, detected the scent of Clark and his guide, beat its chest, and disappeared into the bush"[3].

Chemosit (Nandi Bear, Kerit or Duba (East Africa))

In east African folklore, the Nandi bear is a creature said to live in East Africa.[1][2] It takes its name from the Nandi people who live in western Kenya, in the area the Nandi Bear is reported from. It is also known as Chemosit, Kerit, Koddoelo, Ngoloko, or Duba (which derives from the Arabic words dubb or d.abʕ / d.abuʕ for 'bear' and 'hyena' respectively[4].

Chuchuna

The Chuchunya (also called Chucunaa, Tjutjuna or Siberian Snowman) is hominid cryptid reported to exist in Siberia.

In 1933, Professor P. Dravert became incensed when he heard reports that these creatures were being hunted and petitioned the Soviet government to put an end to this heinous act, stating that Chuchunaa were also citizens of the Soviet Union, and therefore deserved equal protection under the law. Obviously, the Soviet government at the time had no interest in such things. His plea went unheeded[5].

[2] https://en.wikipedia.org/wiki/Almas_(folklore)
[3] https://cryptidarchives.fandom.com/wiki/Beruang_rambai
[4] https://en.wikipedia.org/wiki/Nandi_bear
[5] https://cryptidz.fandom.com/wiki/Chuchunya

Dev
Wildman of west and central Asia, height 4ft10in. Covered in shaggy reddish brown or black hair. Black skin and has horns. Bipedal travels singularly or in pairs. Adult Dev was reported captured in 1933 at a flour mill a few miles from Tutkaul, Tajikistan. Some reports simply suggest it's a devil or demon not a true cryptid.

Geresun Bambursh (Also Kumshin Gerusgu/ Zerleg Khün/ (Mongolian))
The Geresun Bambursh is a cryptid found in Mongolian folklore and literally translated means 'Wild men'[6].

Zerleg Khün is one of the various names for the Kühn Görüessü mongolian 'man-beast'

Variant names hün garees, hün göröös, hün har göröös, khün görüessü 'black man beast', Kümün görüessü, Zerleg Khün. A Wildman skin found preserved in a Mongolian temple proved to belong to a bear.

Harrum-Mo (Central Asia)
The Harrum-mo is an unusual kind of hominid, because it is said to have the power of speech. It also appears to have invented the bow and arrow. It is to be found in the Lunak Valley, Nepal, but no reports have come in since the 19th Century[7].

Jungli-Admi (Sogpa (India))
Another mysterious primate from India. The Sogpa is said to be a beast that likes very high elevations and comes down to the lower valley only in cold weather. It is described as being covered in long yellowish-brown hair, including its face and to stand about 4ft tall, being as comfortable on the ground as in the treetops[8].

Kaftar
Kaftar (كفتار) is Persian for "hyena" and refers to a mystical race of shapeshifters sighted around India's capital of New Delhi.

A medical treatise written in Persian around AD 1376 in Delhi gives detailed prescriptions of a magical nature on how to deal with a man who can transform himself into a striped hyena. This demoniacal being, "half-man, half-hyena," is called kaftar and has the habit of attacking and killing children[9].

---

[6] https://archive.org/stream/landlamasnotesa01rockgoog/landlamasnotesa01rockgoog_djvu.txt
[7] http://cfz-usa.blogspot.com/2021/02/harrum-mo.html?m=0
[8] http://www.unexplainedmysteries.net/s/sogpa.htm
[9] https://cryptidz.fandom.com/wiki/Kaftar

Kiik-Adam

In the Altai Mountain of Central Asia, there are reports of a Bigfoot creature, known by many names, such as Kiyik-Adam, Kishi-Kiyik & Kiik-Kish. In 1911, a Russian zoologist heard of the creature from Kazakh herders, and in 1948, a herder showed a geologist a supposed hand of the Kiyik-Adam[10].

Metch-Kangmi (Yeti)

The term "abominable snowman" originated in a mistranslation of the name meh-toh in 1921. During an expedition to Everest in that year, Sherpa guides working for Lieutenant Colonel C. K. Howard-Bury identified some large footprints as those of a "metoh-kangmi," as Howard-Bury wrote it. This was composed of two Sherpa words, "meh-toh" ("manlike thing that is not a man" or "man-sized wild creature") and "kang-mi," ("snow creature") and was only a generic term. When journalist Henry Newman wrote up the story for the Calcutta Statesman, he rendered the name "metch kangmi" and stated that it was a Tibetan word meaning "abominable snowman" in reference to the creature's purported strong smell[11].

Mo Mo (Marzolf Hill)

Marzolf Hill, a small stone's throw from the end of Allen Street in Missouri USA, where the legend of Mo Mo, the Missouri Monster, emerged in July 1972.

On July 11, 1972, 15-year-old Doris Harrison watched her brothers, ages 5 and 8, play outside as she cleaned a bathroom sink. Startled by her brothers screaming, Doris looked out the window and saw a massive, hairy creature at the edge of the woods.....

Read the full story here: https://missourilife.com/the-legend-of-mo-mo-the-missouri -bigfoot/

M'toto

Toto (1931–1968) (a.k.a. M'Toto meaning "Little Child" in Swahili) was a gorilla that was adopted and raised very much like a human child.

A. Maria Hoyt adopted the baby female gorilla orphaned by a hunt in French Equatorial Africa in 1931. Mrs. Hoyt's husband killed the baby gorilla's father for a museum piece, and his guides killed its mother for fun. Mrs. Hoyt moved to Cuba to provide a more tropical home for Toto. At the age of four or five, Toto adopted a

---

[10] https://www.tumblr.com/cryptid-quest/187540144344/cryptid-of-the-day-kiyik-adam-description-in-the
[11] https://cryptidarchives.fandom.com/wiki/Yeti

kitten named Principe, carrying the kitten with her everywhere. When Toto became too difficult to manage for a private keeper, she was leased to the Ringling Brothers and Barnum and Bailey Circus as a potential mate for another gorilla, Gargantua, a.k.a. Buddy. Toto died in 1968. Toto is buried at "Sandy Lane" Kennels Pet Cemetery in Sarasota, Florida

Peevi (Chinese)
A vaguely described bipedal creature found in ancient Chinese texts; the oldest record is from an encyclopaedia dating back to the 3rd century BC, where it was said to resemble a bear with a longer head and feet, a yellowish coat with white stripes or spots, the strength to knock down trees, and a fierce temper.

Rakshasa (India)
Are a race of usually malevolent beings prominently featured in Hindu mythology. They reside on Earth but possess supernatural powers, which they usually use for evil acts such as disrupting Vedic sacrifices or eating humans.

Rakshi-Bompo (Central Asia)
Alternate name for the Yeti of Central Asia. Tibetan/Indo-Aryan hybrid word meaning 'powerful demon' according to Gordon Creighton. Rakshasa are Hindu demons from the Ramayana, in some old epics they were pre-Aryan inhabitants of India[12].

Shaitan (Middle East)
Shaytan is an evil spirit in Islam, inciting humans (and the jinn) to sin by "whispering" وَسْوَسَة, (waswasa) in their hearts. Although invisible to humans, they are imagined to be ugly and grotesque creatures created from fire[13].

Snally-Gaster (USA)
In American folklore, the snallygaster is a bird-reptile chimera originating in the superstitions of early German immigrants later combined with sensationalistic newspaper reports of the monster. Early sightings associate the snallygaster with Frederick County, Maryland, especially the areas of South Mountain and the Middletown Valley. Later reports would expand on sightings encompassing an area to include Central Maryland and the Washington, DC, metro area[14].

---

[12] https://books.google.co.uk/books?id=z9gMsCUtCZUC&pg=PA456&lpg=PA456&dq=Rakshi-Bompo&source=bl&ots=JWYmiooe5X&sig=ACfU3U1LuSp5vtTBjw_j_gGBYezf-rzHrw&hl=en&sa=X&ved=2ahUKEwjB-6-uvYv-AhXGwKQKHQZFB28Q6AF6BAgTEAM#v=onepage&q=Rakshi-Bompo&f=false
[13] https://en.wikipedia.org/wiki/Shaitan
[14] https://en.wikipedia.org/wiki/Snallygaster

The Skunk Ape

The skunk ape is a cryptic ape-like creature alleged by cryptozoologists to inhabit forests and swamps in the south-eastern United States. Perhaps most prominent in the state of Florida, the alleged creature is also commonly referred to as the Florida Bigfoot, and is often compared to, synonymous with, or called the "cousin" of Bigfoot, a prominent subject within North American popular culture[15].

It was first spotted in February 1970[16] in central Arkansas. It is a North American ape type that is chimpanzee like, more recent sightings in Florida suggest knuckle walking and it gets its name from its intensely unpleasant smell.

Thought to be either from escaped zoo or circus primates.

Urayuli or Hairy Man

Urayuli, or "Hairy Men", are a race of creatures that live in the woodland areas of southwestern Alaska. Stories of the Urayuli describe them as standing 10 feet tall with long shaggy fur and luminescent eyes. They are said to emit a high-pitched cry, resembling that of a loon. Their long, lanky arms have been described as reaching down to their ankles.

It is said the Urayuli are transformed children who become lost in the woods at night. It is possible that this tale was started to keep children indoors at night[17].

Wulgaru (Australia)

It is a living man, but it does not breathe; it is a devil-devil, but it has the form of a man; it is only the trunk of a tree and a few river stones, but it has life and movement. It is Wulgaru, the devil-devil which still lives, and will live forever, the devil-devil which kills men and women and children who break the tribal laws, but which has never been known to touch those who obey[18].

Yeti (Osodrashin/Sandja (Tibetan))

Yeti, also known as the "Abominable Snowman," is a legendary creature said to inhabit the snowy, mountainous regions of the Himalayas in Nepal, Bhutan, and Tibet. The creature is typically described as a large, ape-like creature with black, brown or reddish fur, long arms, and a domed forehead.

---

[15] https://en.wikipedia.org/wiki/Skunk_ape
[16] RF - Skunk ape reports go back at least as far as the 1930s. They did not start in 1970.
[17] https://en.wikipedia.org/wiki/Urayuli
[18] https://www.artistwd.com/joyzine/australia/dreaming/wooden.php

The Yeti is an important part of the folklore and mythology of the Himalayan region, and it has been the subject of numerous sightings and alleged evidence over the years. However, no conclusive and verifiable evidence of the existence of the Yeti has been found, and the scientific community remains sceptical of its existence.

In recent years, some researchers have suggested that the Yeti might be a misidentification of known animals, such as the Tibetan blue bear or the Himalayan brown bear[19], or a cultural interpretation of natural phenomena such as avalanches or rock formations. Despite these theories, the Yeti remains a popular subject of interest and debate among cryptozoologists, adventurers, and folklore enthusiasts.

---

[19] The yeti is not a bear, this is a myth I am trying to quash. Bears walk on all fours and only stand erect for short periods. The yeti walk erect most of the time. Bears have scapula (shoulder blades) that lie flat against their sides as they are quadrupeds. The yeti, like man, has scapula that jut outwards, as it is a biped. That gives it the broad-shouldered appearance. The yeti has a flat, gorilla-like face whereas bears have a long, dog-like snout. The yeti has opposable thumbs and can throw rocks and swing clubs. Bears cannot do this. Native people and witnesses quite rightly become annoyed when westerners say that the yeti is a bear.

Anthropo-Zoology

In her introduction to this book, Odette talks about the 'direct and indirect clues to humanity's start' and throughout the book you will have seen mentions of the word 'Anthropo-Zoology' and how she was keen for this word to become commonplace when referring to the possible link on the tree of life between humans and bigfoot (in his/her many and varied forms).

Odette believed that the theory of a 'missing link' between bigfoot and humans had started to be taken seriously by a very select few individuals, but nothing had been seriously documented or discussed by the time she wrote this book in 1986. It has been the work of 'Fringe science' to investigate unexplained phenomena for centuries and she longed for this type of research to become more mainstream. She quite beautifully referenced Rudyard Kipling when describing what most refer to as jokes and hoaxes as "Something lost behind the ranges" and sadly that is how it still stands.

As Odette mentioned, there was only one serious public investigation she knew of in 1986 when writing the book and this was in 1978 when the University of British Columbia organised a conference on the subject in Vancouver which she attended as listed observer. She mentions it again in later chapters, but few records exist on the conference itself.

Odette rather cleverly discovered a very early example of a missing link when she references a traveller called Edward Tyson in chapter seven of the book. Edward Tyson was one of the first people to observe and write about the Chimpanzee back in 1699, which were once just tales recounted by sailors and often disbelieved. Tyson was an objective traveller and was researching what was once referred to as (a now very outdated term) 'Pygmie' when he spoke of his surprise at the resemblances between chimpanzee and man saying "Our Pygmie is no Man, nor a Common Ape, but a sort of animal between both."

Edward Tyson was observing the Chimpanzee, now of course completely recognised as a species in its own right. Certainly not just a sailor's story or an unexplained beast hiding in the trees. Perhaps this is where research is currently sat with bigfoot and we just need to find them and give them their rightful place, their own 'branch' to sit on.

Odette believed (and many others do too) that this link in human history to bigfoot has been either ignored or covered up for a very long time and she wanted to bring it out into the light of day, have it investigated, researched, and discussed as commonly as one would with any other branches of the 'tree' but was unfortunately unsuccessful before her passing in 1992.

She felt, and I would agree, that it would take a large amount of resource to carry out such investigations by way of surveys, field work, equipment to test and study and that it may face obstacles and objections both financially and politically (who would foot the bill for this worldwide research? Do we all need to write to Elon Musk?), but what would the payoff be? Would we discover a very important missing link in our human history, could it explain things that are currently left unexplainable in those many varied branches of our historical 'tree'?

I'm comforted in the knowledge that she believed it was possible for science to 'come around' to the idea and that what sparse evidence has been found has been documented and shared amongst like-minded folk and she hoped that like herself, people would have a flexible view on the subject, exchange views openly and share information so that the truth could be found.

I believe Odette would have been thrilled by the research carried out in a more modern era by the likes of The Centre for Fortean Zoology and the myriad televised documentaries and series on investigating bigfoot and I would love to be able to bring her dream to life. To have 'Anthropo-Zoology' brought into the mainstream, make it a common word in this area of research and (in the long run) see the research carried out into the links between humanity and the unexplained phenomena known as bigfoot.

Odette says in her final chapter: "Myth began in the long past from a grain of truth, in an age forgotten moment in time. Though I have called the Yeti/Wild Hairy Man Anthropo-zoology, it might be that someday this name I have coined will be accepted as fact".

Odette, I sincerely hope so.

The differences between the various forms of Anthropology

What is Anthropology?
Anthropology is the scientific study of human societies, cultures, and their development. It is a social science that aims to understand the complexities of human behaviour and its relationship with social, political, economic, and cultural structures. Anthropologists seek to answer questions about what it means to be human by studying topics such as social organization, kinship, language, religion, art, and other cultural expressions.

There are four main subfields of anthropology: cultural anthropology, which studies the beliefs, practices, and customs of different cultures; linguistic anthropology, which focuses on the role of language in shaping culture and society; biological (or physical) anthropology, which studies human biological diversity and evolution; and archaeology, which studies human societies and cultures through the analysis of material remains. Together, these subfields provide a comprehensive understanding of human behaviour and social organization, past and present.

Cultural anthropology
Cultural anthropology is a branch of anthropology that studies the cultural behaviours, beliefs, and practices of human societies around the world. It seeks to understand the diversity of human cultures and the ways in which different cultures interact with each other. Cultural anthropologists often conduct fieldwork, which involves living among and observing the daily lives of the people they study. They may also conduct interviews, collect artifacts, and analyse cultural practices and beliefs in order to gain a deep understanding of the cultural systems of a particular society. Cultural anthropology also seeks to address issues of power, inequality, and social justice by examining how cultural practices and beliefs are shaped by historical, political, and economic factors.

Linguistic anthropology
Linguistic anthropology is a subfield of anthropology that focuses on the study of language in its social and cultural context. Linguistic anthropologists investigate how language shapes and is shaped by human societies, cultural norms, and power relations. They examine how people use language to communicate, express social identity, and construct meaning in different cultural contexts.

Linguistic anthropologists often use ethnographic research methods, such as participant observation and interviews, to study language use in specific cultural contexts. They may also use linguistic analysis to understand how language reflects cultural beliefs and practices, and how it changes over time.

Some of the key topics that linguistic anthropologists may study include language acquisition and socialization, multilingualism and language contact, language and identity, language and power, language and gender, and language and technology. They may also work in collaboration with other anthropologists, linguists, and scholars from related fields to explore these topics from interdisciplinary perspectives.

Biological or physical anthropology

Biological anthropology, also known as physical anthropology, is a subfield of anthropology that focuses on the biological aspects of human beings, both past and present. Biological anthropologists study the evolution, anatomy, genetics, and behaviour of humans and their non-human primate relatives.

One of the key areas of focus for biological anthropologists is human evolution. They study the fossil record to understand the origins and evolutionary history of humans, including the emergence of bipedalism, the development of larger brains, and the evolution of language and culture. They may also use genetic analysis to trace human migration patterns and understand the genetic basis of human variation.

Biological anthropologists also study the anatomy and physiology of humans and their primate relatives to better understand the biological basis of human behaviour and adaptation. This may include studying the structure and function of the brain, the mechanics of movement, and the biological basis of diseases and health.

Other topics that biological anthropologists may investigate include the biological and cultural aspects of reproduction, the impact of the environment on human biology and health, and the role of genetics in human development and disease. They may also work in collaboration with other scientists, including geneticists, physiologists, and medical professionals, to better understand the complex relationship between biology, culture, and health.

Archaeological anthropology

Archaeological anthropology, also known as archaeology, is a subfield of anthropology that focuses on the study of human societies and cultures through the analysis of material remains. Archaeological anthropologists study the physical artifacts left behind by past human societies in order to understand their social, economic, and political systems.

Archaeological anthropology is concerned with the reconstruction of past human behaviour and societies. This may involve excavating archaeological sites, analysing artifacts and other material remains, and using a variety of scientific techniques to date and interpret the data.

One of the key areas of focus for archaeological anthropology is the study of human cultural evolution over time. This may involve studying the development of agriculture, the rise of complex societies, and the emergence of technological innovations such as writing and metallurgy. Archaeological anthropology also investigates the relationships between different cultures and societies, including trade, migration, and conflict.

Other topics that archaeological anthropologists may investigate include the social and economic organization of ancient societies, the impact of environmental change on human societies, and the ways in which past societies dealt with issues such as disease and population growth. They may also work in collaboration with other anthropologists, historians, and scientists to better understand the complexities of human cultural evolution and change over time.

Social anthropology
Social anthropology, also known as cultural anthropology in some contexts, is a subfield of anthropology that focuses on the study of human societies and cultures in their social, political, and economic contexts. Social anthropologists study the ways in which people interact with each other and with their environment, and how these interactions are shaped by cultural and social norms.

Social anthropologists often conduct fieldwork, which involves living among and observing the daily lives of the people they study. They may also conduct interviews, collect artifacts, and analyse cultural practices and beliefs in order to gain a deep understanding of the cultural systems of a particular society.

One of the key areas of focus for social anthropology is the study of social organization and structure. This may involve investigating the ways in which people organize themselves into groups, such as families, clans, or tribes, and how they interact with other groups in their society. Social anthropologists also study the ways in which social norms and values shape behaviour, including the roles of gender, race, and ethnicity in society.

Other topics that social anthropologists may investigate include economic systems and exchange, political systems and power relations, religion and spirituality, and the impact of globalization and modernization on traditional societies. They may also work in collaboration with other anthropologists, sociologists, and scholars from related fields to explore these topics from interdisciplinary perspectives.

Anthrozoology

Anthrozoology is a relatively new field of study that explores the relationships between humans and animals. It is an interdisciplinary field that draws on insights from anthropology, psychology, biology, and other related fields.

Anthrozoologists study the ways in which humans and animals interact with each other, and the ways in which these interactions are shaped by cultural, social, and environmental factors. This may involve investigating the historical and cultural relationships between humans and domesticated animals, such as dogs, cats, and livestock.

One of the key areas of focus for anthrozoology is the study of human attitudes and behaviours towards animals, including attitudes towards animal welfare, conservation, and the use of animals in agriculture, entertainment, and scientific research. Anthrozoologists may also investigate the impact of animals on human health and well-being, including the therapeutic benefits of animal-assisted therapy.

Other topics that anthrozoologists may investigate include the role of animals in human cultural and spiritual practices, the ethics of animal rights, and the ways in which humans and animals adapt to each other in changing environmental conditions. They may also work in collaboration with other scientists, animal welfare organizations, and policy makers to better understand the complex relationships between humans and animals in different contexts.

Anthrozoology is a relatively new field of study that emerged in the late 20th century, although its roots can be traced back to earlier traditions of human-animal studies. The term "anthrozoology" was first coined by biologist John Bradshaw in the 1990s, to describe the interdisciplinary study of human-animal relationships.

While the field of anthrozoology is relatively new, humans have had a long-standing relationship with animals throughout history, and there have been many cultural traditions and practices that involve animals in various ways. However, it was not until the 20th century that the study of human-animal relationships began to emerge as a distinct field of inquiry.

Today, anthrozoology is a rapidly growing field that draws on insights from anthropology, psychology, biology, and other related fields, and has important implications for animal welfare, conservation, and human well-being.

STILL ON THE TRACK OF UNKNOWN ANIMALS

T he Centre for Fortean Zoology, or CFZ, is a non profit-making organisation founded in 1992 with the aim of being a clearing house for information, and coordinating research into mystery animals around the world.

We also study out of place animals, rare and aberrant animal behaviour, and Zooform Phenomena; little-understood "things" that appear to be animals, but which are in fact nothing of the sort, and not even alive (at least in the way we understand the term).

Not only are we the biggest organisation of our type in the world, but - or so we like to think - we are the best. We are certainly the only truly global cryptozoological research organisation, and we carry out our investigations using a strictly scientific set of guidelines. We are expanding all the time and looking to recruit new members to help us in our research into mysterious animals and strange creatures across the globe.

Why should you join us? Because, if you are genuinely interested in trying to solve the last great mysteries of Mother Nature, there is nobody better than us with whom to do it.

We publish a journal *Animals & Men*. Each issue contains nearly 100 pages packed with news, articles, letters, research papers, field reports, and even a gossip column! The magazine is Royal Octavo in format with a full colour cover. You also have access to one of the world's largest collections of resource material dealing with cryptozoology and allied disciplines, and people from the CFZ membership regularly take part in fieldwork and expeditions around the world.

The CFZ is managed by a board of trustees, with a non-profit making trust registered with HM Government Stamp Office. The board of trustees is supported by a Permanent Directorate of full and part-time staff, and advised by a Consultancy Board of specialists - many of whom are world-renowned experts in their particular field. We have regional representatives across the UK, the USA, and many other parts of the world, and are affiliated with other organisations whose aims and protocols mirror our own.

You'll find that the people at the CFZ are friendly and approachable. We have a thriving forum on the website which is the hub of an ever-growing electronic community. You will soon find your feet. Many members of the CFZ Permanent Directorate started off as ordinary members, and now work full-time chasing monsters around the world.

Write to us, e-mail us, or telephone us. The list of future projects on the website is not exhaustive. If you have a good idea for an investigation, please tell us. We may well be able to help.

We are always looking for volunteers to join us. If you see a project that interests you, do not hesitate to get in touch with us. Under certain circumstances we can help provide funding for your trip. If you look on the future projects section of the website, you can see some of the projects that we have pencilled in for the next few years.

In 2003 and 2004 we sent three-man expeditions to Sumatra looking for Orang-Pendek - a semi-legendary bipedal ape. The same three went to Mongolia in 2005. All three members started off merely subscribers to the CFZ magazine. Next time it could be you!

We have no magic sources of income. All our funds come from donations, membership fees, and sales of our publications and merchandise. We are always looking for corporate sponsorship, and other sources of revenue. If you have any ideas for fund-raising please let us know. However, unlike other cryptozoological organisations in the past, we do not live in an intellectual ivory tower. We are not afraid to get our hands dirty, and furthermore we are not one of those organisations where the membership have to raise money so that a privileged few can go on expensive foreign trips. Our research teams, both in the UK and abroad, consist of a mixture of experienced and inexperienced personnel. We are truly a community, and work on the premise that the benefits of CFZ membership are open to all.

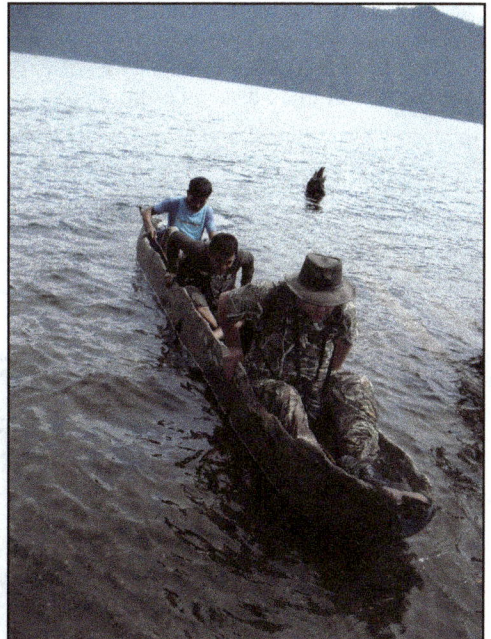

Reports of our investigations are published on our website as soon they are available. Preliminary reports are posted within days of the project finishing.

Each year we publish a 200 page yearbook containing research papers and expedition reports too long to be printed in the journal. We freely circulate our information to anybody who asks for it.

We have a thriving YouTube channel, CFZtv, which has well over two hundred self-made documentaries, lecture appearances, and episodes of our monthly webTV show. We have a daily online magazine, which has over a million hits each year.

From 2000—2016 we held our annual convention - the Weird Weekend. It went on hiatus because of the illness of several of the major personnel and the eventual death of one of them. But we plan to bring it back soon. It is three days of lectures, workshops, and excursions. But most importantly it is a chance for members of the CFZ to meet each other, and to talk with the members of the permanent directorate in a relaxed and informal setting and preferably with a pint of beer in one hand. Since 2006 - the Weird Weekend has been bigger and better and held in the idyllic rural location of Woolsery in North Devon.

Since relocating to North Devon in 2005 we have become ever more closely involved with other community organisations, and we hope that this trend will continue. We have also worked closely with Police Forces across the UK as consultants for animal mutilation cases, and we intend to forge closer links with the coastguard and other community services. We want to work closely with those who regularly travel into the Bristol Channel, so that if the recent trend of exotic animal visitors to our coastal waters continues, we can be out there as soon as possible.

Apart from having been the only Fortean Zoological organisation in the world to have consistently published material on all aspects of the subject for over a decade, we have achieved the following concrete results:

• Disproved the myth relating to the headless so-called sea-serpent carcass of Durgan beach in Cornwall 1975

• Disproved the story of the 1988 puma skull of Lustleigh Cleave

- Carried out the only in-depth research ever into the mythos of the Cornish Owlman.
- Made the first records of a tropical species of lamprey
- Made the first records of a luminous cave gnat larva in Thailand
- Discovered a possible new species of British mammal - the beech marten
- In 1994-6 carried out the first archival fortean zoological survey of Hong Kong
- In the year 2000, CFZ theories were confirmed when a new species of lizard was added to the British List
- Identified the monster of Martin Mere in Lancashire as a giant wels catfish
- Expanded the known range of Armitage's skink in the Gambia by 80%
- Obtained photographic evidence of the remains of Europe's largest known pike
- Carried out the first ever in-depth study of the ninki-nanka
- Carried out the first attempt to breed Puerto Rican cave snails in captivity
- Were the first European explorers to visit the `lost valley` in Sumatra
- Published the first ever evidence for a new tribe of pygmies in Guyana
- Published the first evidence for a new species of caiman in Guyana
- Filmed unknown creatures on a monster-haunted lake in Ireland for the first time

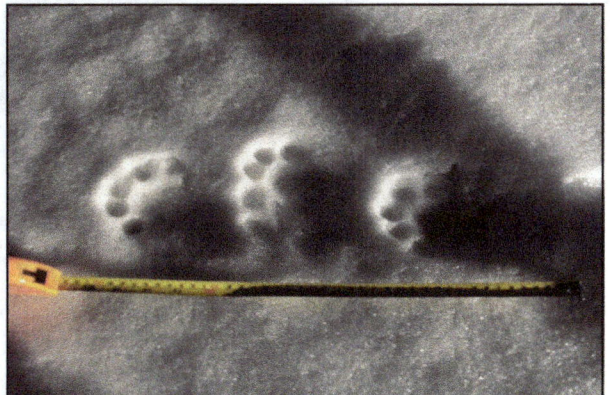

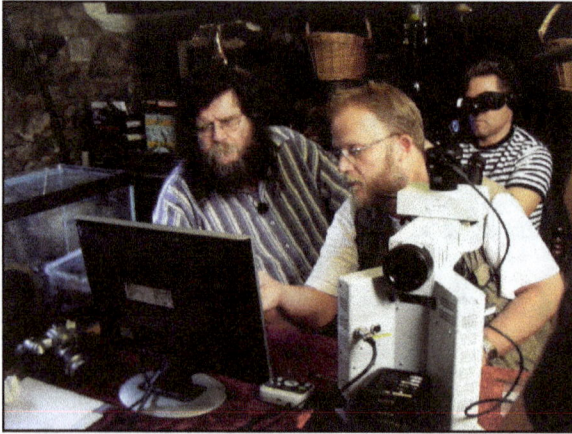

Alive...The Missing Link

- ~~Had a sighting of orang pendek~~ in Sumatra in 2009
- Found leopard hair, subsequently identified by DNA analysis, from rural North Dev on in 2010
- Brought back hairs which ap pear to be from an unknown primate in Sumatra
- Published some of the best evidence ever for the almasty in southern Russia

CFZ Expeditions and Investigations include:

- 1998 Puerto Rico, Florida, Mexico (Chupacabras)
- 1999 Nevada (Bigfoot)
- 2000 Thailand (Naga)
- 2002 Martin Mere (Giant catfish)
- 2002 Cleveland (Wallaby mutilation)
- 2003 Bolam Lake (BHM Reports)

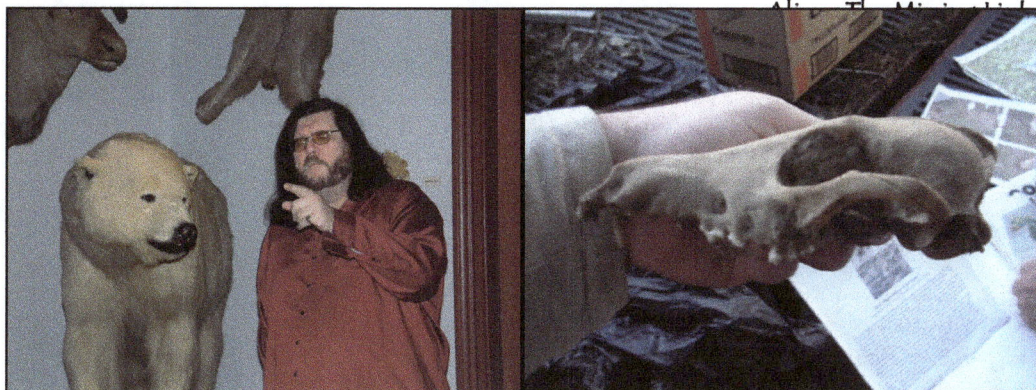

- 2003 Sumatra (Orang Pendek)
- 2003 Texas (Bigfoot; giant snapping turtles)
- 2004 Sumatra (Orang Pendek; cigau, a sabre-toothed cat)
- 2004 Illinois (Black panthers; cicada swarm)
- 2004 Texas (Mystery blue dog)
- Loch Morar (Monster)
- 2004 Puerto Rico (Chupacabras; carnivorous cave snails)
- 2005 Belize (Affiliate expedition for hairy dwarfs)
- 2005 Loch Ness (Monster)
- 2005 Mongolia (Allghoi Khorkhoi aka Mongolian death worm)

- 2006 Gambia (Gambo - Gambian sea monster , Ninki Nanka and Armitage's skink
- 2006 Llangorse Lake (Giant pike, giant eels)
- 2006 Windermere (Giant eels)
- 2007 Coniston Water (Giant eels)
- 2007 Guyana (Giant anaconda, didi, water tiger)
- 2008 Russia (Almasty)
- 2009 Sumatra (Orang pendek)
- 2009 Republic of Ireland (Lake Monster)
- 2010 Texas (Blue Dogs)
- 2010 India (Mande Burung)
- 2011 Sumatra (Orang-pendek)
- 2012 Sumatra (Orang Pendek)
- 2014 Tasmania (Thylacine)
- 2015 Tasmania (Thylacine)
- 2016 Tasmania (Thylacine)
- 2017 Tasmania (Thylacine)
- 2018 Tajikistan (Gul)
- 2020 Forest of Dean (Lynx)

For details of current membership fees, current expeditions and investigations, and voluntary posts within the CFZ that need your help, please do not hesitate to contact us.

The Centre for Fortean Zoology,
Myrtle Cottage,
Woolfardisworthy,
Bideford, North Devon
EX39 5QR

Telephone 01237 431413
Fax+44 (0)7006-074-925
eMail info@cfz.org.uk

Websites:

www.cfz.org.uk
www.weirdweekend.org

ANIMALS & MEN ISSUES 16-20
THE JOURNAL OF THE CENTRE FOR FORTEAN ZOOLOGY
NEW HORIZONS
Edited by Jon Downes

BIG CATS LOOSE IN BRITAIN
MARCUS MATTHEWS

PREDATOR DEATHMATCH
NICK MOLLOY
WITH ILLUSTRATIONS BY ANTHONY WALLIS

THE WORLD'S WEIRDEST PUBLISHING COMPANY

PHENOMENA

Edited by
Jonathan Downes and Richard Freeman

FOREWORD BY Dr. KARL SHUKER

A DAINTY DIARY
Tales from Travels
tropical North

CARL PORTMAN

THE COLLECTED POEMS
Dr Karl P.N. Shuker

STRANGELY STRANGE
ly normal

an anthology of writings by
ANDY ROBERTS

HOW TO START A PUBLISHING EMPIRE

Unlike most mainstream publishers, we have a non-commercial remit, and our mission statement claims that "we publish books because they deserve to be published, not because we think that we can make money out of them". Our motto is the Latin Tag *Pro bona causa facimus* (we do it for good reason), a slogan taken from a children's book *The Case of the Silver Egg* by the late Desmond Skirrow.

WIKIPEDIA: "The first book published was in 1988. *Take this Brother may it Serve you Well* was a guide to Beatles bootlegs by Jonathan Downes. It sold quite well, but was hampered by very poor production values, being photocopied, and held together by a plastic clip binder.

In 1988 A5 clip binders were hard to get hold of, so the publishers took A4 binders and cut them in half with a hacksaw. It now reaches surprisingly high prices second hand.

The production quality improved slightly over the years, and after 1999 all the books produced were ringbound with laminated colour covers. In 2004, however, they signed an agreement with Lightning Source, and all books are now produced perfect bound, with full colour covers."

Until 2010 all our books, the majority of which are/were on the subject of mystery animals and allied disciplines, were published by `CFZ Press`, the publishing arm of the Centre for Fortean Zoology (CFZ), and we urged our readers and followers to draw a discreet veil over the books that we published that were completely off topic to the CFZ.

However, in 2010 we decided that enough was enough and launched a second imprint, `Fortean Words` which aims to cover a wide range of non animal-related esoteric subjects. Other imprints will be launched as and when we feel like it, however the basic ethos of the company remains the same: Our job is to publish books and magazines that we feel are worth publishing, whether or not they are going to sell. Money is, after all - as my dear old Mama once told me - a rather vulgar subject, and she would be rolling in her grave if she thought that her eldest son was somehow in `trade`.

Luckily, so far our tastes have turned out not to be that rarified after all, and we have sold far more books than anyone ever thought that we would, so there is a moral in there somewhere...

Jon Downes,
Woolsery, North Devon
July 2010

CFZ PRESS

CFZ Press is our flagship imprint, featuring a wide range of intelligently written and lavishly illustrated books on cryptozoology and the quirkier aspects of Natural History.

CFZ Classics is a new venture for us. There are many seminal works that are either unavailable today, or not available with the production values which we would like to see. So, following the old adage that if you want to get something done do it yourself, this is exactly what we have done.

Desiderius Erasmus Roterodamus (b. October 18th 1466, d. July 2nd 1536) said: "When I have a little money, I buy books; and if I have any left, I buy food and clothes," and we are much the same. Only, we are in the lucky position of being able to share our books with the wider world. CFZ Classics is a conduit through which we cannot just re-issue titles which we feel still have much to offer the cryptozoological and Fortean research communities of the 21st Century, but we are adding footnotes, supplementary essays, and other material where we deem it appropriate.

http://www.cfzpublishing.co.uk/

Fortean Words is a new venture for us. The F in CFZ stands for "Fortean", after the pioneering researcher into anomalous phenomena, Charles Fort. Our Fortean Words imprint covers a whole spectrum of arcane subjects from UFOs and the paranormal to folklore and urban legends. Our authors include such Fortean luminaries as Nick Redfern, Andy Roberts, and Paul Screeton. . New authors tackling new subjects will always be encouraged, and we hope that our books will continue to be as ground-breaking and popular as ever.

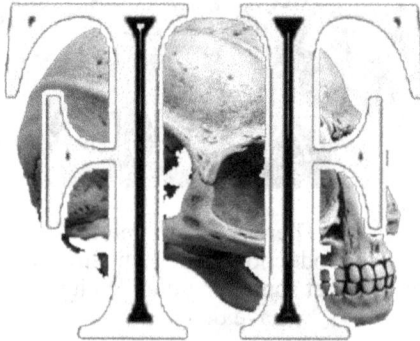

Just before Christmas 2011, we launched our third imprint, this time dedicated to - let's see if you guessed it from the title - fictional books with a Fortean or cryptozoological theme. We have published a few fictional books in the past, but now think that because of our rising reputation as publishers of quality Forteana, that a dedicated fiction imprint was the order of the day.

http://www.cfzpublishing.co.uk/

www.ingramcontent.com/pod-product-compliance
Lightning Source LLC
Chambersburg PA
CBHW072143270326
41931CB00010B/1868